Mountain Trails

MOUNTAIN TRAILS LOGO COURTESY OF WESTERN PRODUCER

Mountain Trails

Memoirs of an Alberta Forest Ranger in the Mountains and Foothills of the Athabasca Forest 1920–1945

by Jack Glen Sr.

Edited by
Robert Mueller, Peter Murphy and Bob Udell

With Bruce Mayer and Bob Stevenson

Foreword by Jim Glen
Introduction by Peter Murphy

Dedicated to the Memory of Jack Glen 1891–1983
and his son Jim Glen 1931–2007

Published by
Foothills Research Institute and
Alberta Department of Sustainable
Resource Development

SPONSORS

Foothills Research Institute
P.O. Box 6330
1176 Switzer Drive
Hinton AB
T7V 1X6
Canada
1-780-865-8330
www.foothillsresearchinstitute.ca

Information Centre
Alberta Sustainable
Resource Development
Main Floor, 9920 108 Street
Edmonton Alberta Canada
T5K 2M4
1-877-944-0313
www.srd.gov.ab.ca

The publishers gratefully acknowledge the financial support of the Forest Resource Improvement Association of Alberta and The Forest History Association of Alberta.

Cover and interior design by John Luckhurst

Maps by Foothills Research Institute

Front cover: Jack Glen Photos, Alberta Forest History Photographic Collection.

Printed in Canada by Friesens

2008 / 1

Library and Archives Canada Cataloguing in Publication

Glen, Jack, 1891-1983
Mountain trails: memoirs of an Alberta forest ranger in the mountains and foothills of the Athabasca forest, 1920-1945 / by Jack Glen; edited by Robert Mueller, Peter Murphy and Bob Udell; with Bruce Mayer and Bob Stevenson; foreword by Jim Glen; introduction by Peter Murphy.

"Dedicated to the memory of Jack Glen, 1891-1983, and his son Jim Glen, 1931-2007."
Includes bibliographical references and index.
ISBN 978-1-896585-13-0 (bound)
ISBN 978-1-896585-12-3 (pbk.)

1. Glen, Jack, 1891-1983.
2. Athabasca Forest Reserve (Alta.)–History.
3. Entrance (Alta.)–History.
4. Rocky Mountains Forest Reserve (Alta.)–History.
5. Forests and forestry–Alberta–History.
6. Forest management–Alberta–History.
7. Forest rangers–Alberta–Biography.
I. Mueller, Robert, 1967- II. Murphy, Peter John, 1930- III. Udell, Robert, 1943- IV. Mayer, Bruce V. Stevenson, Bob VI. Alberta. Alberta Sustainable Resource Development VII. Foothills Research Institute VIII. Title.

SD129.G56A3 2008 634.9092 C2008-905657-4

Contents

FOREWORD vii

PREFACE AND ACKNOWLEDGEMENTS ix
- Notes on the Author xi

INTRODUCTION: THE HISTORICAL CONTEXT 1
- Prehistory and Fur Trade 4
- National Parks and Forest Reserves 6
- Entrance 7

CHAPTER ONE: FIRST DAYS ON THE TRAIL 9
- From Mountie to Ranger: A New Career in the Athabasca Forest Reserve 10
- My New Frontier Home 14
- First Patrol into the Athabasca Forest Reserve 18
- A New Assistant, Building the Eagles Nest Cabin 27
- A Long Foot Patrol 31

CHAPTER TWO: TRAVELS WITH FRENCHY RIVIERE 45
- On the Trail into Jasper National Park and the Athabasca Forest Reserve 48
- North on the Lower Trail 64
- Fire at Rock Lake 71

CHAPTER THREE: ODD JOBS AND FAILED ENTERPRISES, 1921–1923 77
- A Poor Farming Prospect 78
- Back to the Rocky Mountain Forest Reserve – Fire on the Berland 79
- Back on Patrol 84
- A Short Career in Logging and Sawmilling, Winter 1922–1923 86
- The End of My Logging Career 93

CHAPTER FOUR: BACK ON THE TRAIL, 1923–1924 95
- Hunting and Reconnoitring to the Smoky River and Sheep Creek 104
- Winter at Entrance Headquarters, 1923–1924 115
- Building the Athabasca Lookout, 1924 120
- A New Trail Season Begins, 1924 124
- A Busy Winter in Entrance, 1925 134

CHAPTER FIVE: A RENEWED TRAIL PROGRAM, 1925–1927 141
- Berland River Trail 142
- Back to the Mountains – Adams Creek and Kvass Trails 146
- A Busy Fall and Winter 153
- Wildhay–Moberly, À La Pêche and Muskeg Trails, 1926 157
- Hardships on the Rocky Pass Trail 162
- From Rail to Road, 1927–28 168

Building the Sheep Creek Trail, 1927 ... 171
Finishing the Smoky River Trail to Grand Cache, 1927 ... 186

CHAPTER SIX: HARD TIMES COMING, 1928–1929 ... 197
Finishing the Eagles Nest Cabin ... 198
Last Patrol, 1928 ... 202
Patrols and Cabins with Bill Smith, 1929 ... 206
Fire on the Smoky River ... 214
Harry Phillips and the Wounded Grizzly ... 221

CHAPTER SEVEN: CHANGED CIRCUMSTANCES, 1930–1931 ... 225
Cabins, Messy Campers and Treed by a Bear ... 227
Youth and Inexperience Yield Tragic Results ... 235

CHAPTER EIGHT: FIREFIGHTING AND SEASONAL WORK ... 239
Smoky River Fire ... 240
A Stubborn Fire on the Mountain Trail ... 247
The Sheep Hunt ... 254
Another Season on the Trail, 1934 ... 256
Fire in Rock Creek Valley ... 260
Fred Wharton's Story ... 263

CHAPTER NINE: LAST TRAILS, 1935–1945 ... 267
Family Life in Entrance ... 268
National Forestry Program, 1938 ... 269
The War Years ... 269
Ted Hammer Leaves the Athabasca Forest, 1940 ... 271
The Hockey Player ... 274
Serious Infection on Patrol ... 276
The First School at Entrance ... 279
Fire at the Drinnan Mine ... 282
Patrols and Bear Encounters, 1941 ... 282
Promotion and Last Patrol ... 288

APPENDIX 1: A Letter from Jack Glen to John Currat, September 25, 1971 ... 289

APPENDIX 2: Jack Glen, Forest Ranger, History of Employment, 1920–1945 ... 292

APPENDIX 3: Local friends and neighbours attending Jack Glen's farewell party, Old Entrance, 28 February 1945 ... 294

APPENDIX 4: Rangers and Service Years, Athabasca Forest, 1914-1945 ... 295

Foreword

I was fortunate to spend my early years in Entrance, Alberta, where my father, Jack Glen, lived and worked as an Alberta Forest Service ranger on the Athabasca Forest Reserve. One of my earliest memories is working in the blacksmith shop with him while he was shoeing Frenchie and Marie, his team of Belgian horses. I was five or six at the time, and my job was to turn the blower on the forge as he shaped the shoes to fit the shape of the horses' hoofs. It was around that time that he made me a rocking horse for Christmas. This was a real horse to me and I put lots of miles and adventures on him in place of being able to ride with my father on his patrols.

My mother needed help at home during my father's long absences on patrol; sometimes he was gone for a month at a time. My father taught me how to look after the garden, as we depended on it a lot for our food. I enjoyed working and spending time with him and learning to help my mother with the weeding. He built a root cellar near the house for potatoes and other root vegetables and my mother canned other food and meat – mainly Rocky Mountain bighorn sheep – for the winter months.

In the summer, when my father was out on patrol for long periods of time, we would find out when he was returning and my mother and I would walk down the railroad tracks (which later became the first road to Brûlé) to meet

Jim Glen and his mother at their Entrance home. Jim is sitting on Glen's horse, as he describes in this Foreword. Late 1930s.

JACK GLEN PHOTOS, ALBERTA FOREST HISTORY PHOTOGRAPHIC COLLECTION

him. He would put me up on the saddle and he and mother would walk together while I rode his saddle horse, which was a lot more fun to ride than my rocking horse. Often these visits were very short as he restocked his supplies to head back out on his patrol.

By the time I was twelve or thirteen, I was old enough to help my father with the fall haying if the other ranger was busy with some other duty. One time the team got out of the pasture and my father let me ride his saddle horse along with another ranger, Carl Larson, to find them. At this time, he also taught me to drive a 1939 Chev truck with a floor shift. When we moved to Terrace some years later, the policeman testing me for my licence asked where I learned to drive so well and I told him in the hay fields of Alberta.

We saw more of my father in the winter time and I greatly enjoyed the times we spent together skating in front of our house on the Athabasca River, where he cleaned off a patch of ice for this purpose.

To this day, I still think of the things my father taught me. He was truly an outdoors man, gentle and kind. Although small in stature, he was a strong man who won the respect of all he met. He lived a truly adventurous life, but I think what made him the man he was, was not so much what he did as it was what he believed. The hills, mountains, rivers, lakes and mountain streams all shaped him and were a real part of his life. His kindness and love for my mother and myself has influenced me in how I treat my own family with respect and honour. His view of the land was that it was there to be enjoyed but not destroyed and I believe that still.

Jim Glen
Cardston, Alberta
January 29, 2006

Preface and Acknowledgements

Jack Glen was a forest ranger at Entrance from 1920 to 1945, starting in Dominion Forestry Branch (DFB) days and transferring smoothly to the Alberta Forest Service (AFS) when it assumed control of the forests of Alberta in 1930 with the Transfer of Resources Act. During this time he kept a detailed memoir of his experiences, and this memoir became the basis for his book "Mountain Trails." He could not find a publisher for his book, but in 1969 the Western Producer agreed to produce a serialized version of the memoir in its magazine, starting June 5, 1969.

Mrs. Sarah Nicole McCreedy was Forest Service librarian at the time. She saw the articles, cut and pasted them and had them photocopied. Peter Murphy and Bob Stevenson came across a set of the originals in the history storage room at the Environmental Training Centre in the fall of 1997, the copy had been saved and looked after by Terry Smith of the Forest Technology School. The text was subsequently transferred to a Word file for ongoing reference by Peter Murphy and later by Fran Hanington of the Foothills Model Forest[1].

In late 2002 Bob Udell, program lead of the Adaptive Forest Management/History Project at Foothills Model Forest, contacted Jack Glen's son, Jim, of Terrace, B.C., seeking permission to reproduce the series as part of the model forest series, working from the Western Producer serialization. Mr. Glen not only gave permission but also provided a copy of the original unedited manuscript, which the team used instead. The Glen family also donated Jack Glen's forestry photos to the Forest Protection Historical Photographic Collection, many of which appear in this book. Further, Mr. Glen wrote a Foreword that sets a personal context. He was very proud of his father's achievements and gave us great support. He was looking forward to publication but unfortunately passed away on September 23, 2007, before he was able to savour the book. We have dedicated this to the memory of both Jim and his father.

In the 1970s Jack Wright of North Western Pulp and Power Limited (now West Fraser Mills, Hinton Wood Products) located and mapped all the AFS patrol trails in the NWP&P forest. These records, as well as original DFB/AFS maps, were used and adapted by Bob Udell with the help of the GIS team at Foothills Research Institute — Debbie Mucha, Heather Daw, and Peter Caputa — to produce the basemaps used in this book. Joan Udell added her considerable skills to hand letter details and names for the final maps.

In producing this transcription, the team strove to remain true to the original manuscript with editing to improve readability, context and chronology of events. We are indebted to editor Fran Aitkens who provided thoughtful and professional advice to improve the flow and logic of the text. Few dates were

[1] In early 2008, Foothills Model Forest changed its name to Foothills Research Institute to better represent its expanding partnership and sustainable forest management research program.

provided in the original manuscript, so most are interpreted from the context of the times described, or in some cases from Forest Service employment records where the service times of Mr. Glen and others are found. Any errors in reporting dates are ours alone, for which we apologize. This is the third book in the Model Forest's History Series that has been designed by John Luckhurst, and we greatly appreciate his tasteful and skilled artistry that adds such colour and interest to the story.

Various sources were consulted to provide additional details to supplement the text of Mr. Glen's memoir. These included Jack's own story of his early days in Canada, *Where the Rivers Meet* (John Glen Sr., New Rapier Press, 1977); Peter J. Murphy *et al's A Hard Road to Travel* (Foothills Model Forest and The Forest History Society, 2007); James G. MacGregor's biography of Shand Harvey, *Pack Saddles to Tête Jaune Cache* (Hurtig Publishers, 1962); Hazel Hart's *History of Hinton* (Friesen, 1980); Richard Wuorinen's *History of Grande Cache* (Grande Cache Historical Society, 1997); Susan Feddema-Leonard's *People and Peaks of the Willmore* (Willmore Wilderness Foundation and Whitefox Circle Inc., 2007), Neil Gilliat's *If Moose Could Only Talk* (Brightest Pebble Publishing Company, Inc. 1998), Aphrodite Karamitsanis' *Place Names of Alberta: Mountains, Mountain Parks and Foothills* (University of Calgary Press, 1991), Dr. William C. Taylor's biography of Curly Phillips, *Tracks Across My Trail* (Jasper-Yellowhead Historical Society, 1984); and Professor I.S. MacLaren's research files amassed in support of his many historic research projects. We especially appreciate the assistance of Mary Luger of Entrance. Granddaughter of Roy Woodley, she is working on a pictorial history of Entrance and provided editorial corrections, additional background and photographs about Entrance, the Woodley Brothers and Ward Badgley. Lucy McRae, daughter of ranger Ernie Harrison, provided photographs and important insights. As well, we acknowledge with thanks information provided by Jackie Hanington, herself a forest ranger's wife on the Athabasca Forest, and Lena Ouellet and Ron Pelletier about the Métis family histories.

We are also indebted to the following people and organizations who provided photographs or art for use in this book: Jim and Carol Glen (Jack Glen Photos), Western Producer for permission to reproduce its original distinctive chapter heading block, the Alberta Forest History Photographic Collection (including Dominion Forestry Branch, Stan Hughes, Gordon Watt Collections), Foothills Research Institute, Hinton Training Centre, Mary Luger (Roy Woodley Collection) and Cindy LeClercq, Jackie Hanington (her photos as well as Mark Truxler Collection), Lucy McRae (Ernest Harrison Collection),

Hinton Library and Archives (Hazel Hart, Cairns and Moberly Collections), the Glenbow Museum and Archives, the Jasper Yellowhead Museum and Archives, Library and Archives Canada, Parks Canada, Robert Guest, John Hiebert, Brian Carnell, Charles Kay, Bob Stevenson, Bob Udell, Neil Gilliat, the late Dennis Radcliffe, Stan Tietge (Fred Tietge Collection), Tom Vinson Jr., Susan Feddema-

Leonard and her associates from the Willmore Wilderness Foundation including Mac Elder, Fay McReady, Ishbel (Hargreaves) Cochrane, and Dave Simpson, Istockphoto and Hinton Wood Products.

The publishers gratefully acknowledge the financial support provided by the Forest Resource Improvement Association of Alberta, Alberta Sustainable Resource Development and the Forest History Association of Alberta.

Notes on the Author

Jack Glen was born in Scotland in 1891 and emigrated to Canada as a young man, intending to take up cattle ranching in the West. Taking a train to Vancouver, he carried on to Duncan where his brother owned a meat market. He worked for a while at a dairy farm in the area, but soon tired of what he considered the slow pace of life there. His longing for adventure was stimulated by a discussion with a local farmer, Mr. Jeffrey, who was selling his farm and moving his wife and family to the François Lake area of the Bulkley Valley in north-central British Columbia, where the land was rich and available for homesteading. Jack asked to join them in this adventure, and in 1911 they travelled north by boat to Prince Rupert, by riverboat to Hazelton, and eventually by horseback to the François Lake area. There he worked at various jobs including hauling supplies for the construction of the Grand Trunk Pacific Railroad as it passed through on the way to Prince Rupert.

In the early years of the First World War he moved to Wainwright, Alberta, where he tried to enlist in the Canadian Army but was rejected. After a brief attempt at farming, he bought a livery stable which he ran successfully until a recruiting poster for the Royal North-West Mounted Police caught his eye. The RNWMP was recruiting for the war effort, planning to send several squadrons overseas. Jack's sense of adventure and wanderlust was again stimulated and, despite his fears about failing to meet the height requirement (5 feet 8 or 9 inches, as he recalled in his later memoir), he was accepted and sent to Europe. He emerged from the war unscathed, and after he returned to Canada the RNWMP assigned him to Winnipeg. He participated in the police action at the 1919 Winnipeg Riots, which left a deep impression on him for the rest of his life. He then transferred to Edmonton and served for a short time with "G" Division. In 1920 police duties formerly carried out by the RNWMP were being taken over by the new Alberta Provincial Police, and Jack did not want to continue with a force whose role was greatly diminished. He resigned and joined the Dominion Forestry Branch, transitioning in 1930 to the newly created Alberta Forest Service, where he spent some of his happiest days. It is this period, which is the focus of this book.

When he left the Forest Service in 1945, Jack and family moved to the Terrace area in British Columbia where he worked as a forest protection officer for the Columbia Cellulose Company until retirement. In the late 1960s he visited Peter Murphy and staff of the Forest Technology School in Hinton. He later moved

to Chemainus where he remained until his death in 1983 at age 92. He was well known locally as an artist, writer, inventor and gardener. Mr. Glen published one book, Where the Rivers Meet, the story of his emigration and early years in Canada, with special focus on the development of the Skeena and Bulkley river valleys. His Mountain Trails story was condensed and published by the Western Producer in 1969. The version in this book was expanded, adding additional material from his original text. His last book, G Division, describing his time with the RNWMP, was never published; Jim and Carol Glen have given a copy of the manuscript to the Foothills Research Institute, where it has been transcribed for potential future publication.

Opposite page: Jack Glen at his home in Entrance, late 1930s.

JACK GLEN PHOTOS, ALBERTA FOREST HISTORY PHOTOGRAPHIC COLLECTION

Introduction – The Historical Context

By P. J. Murphy

Jack Glen arrived in Entrance, Alberta, in May 1920 to work as a Dominion forest ranger on the Athabasca Forest. The story that follows is the tale of his work and adventures, seeing the country and its inhabitants through the eyes of a newcomer to the region. There was a lot of "new" in the country at that time and Jack Glen was not the only recent arrival. This was a time of great change that had been slowly gathering momentum over the preceding 110 years, since the area had been remote from settlements and difficult to access.

Jasper Park was established 1907, only thirteen years earlier. The Athabasca Forest of the Rocky Mountains Forest Reserve was first defined in 1910, first staffed in 1912 and not greatly developed until after World War I. The railways were also new – construction of the two competing lines along the upper Athabasca River and through Yellowhead Pass started around 1910. Passenger service on the Grand Trunk Pacific Railway to Jasper began January 1912, the line passing along the south bank of the Athabasca River. The Entrance Station was opened in 1914 when the Canadian Northern Railway crossed to the north side up-river from Prairie Creek. By the time Glen arrived in 1920 the two railways had failed financially, their lines had been combined, and in 1921 they were restructured as the Canadian National Railway.

Advent of the railway ended the area's isolation. It also made it easy to transport food from outside, ending dependency on local hunters and packers to provide meat and supplies. The railway also brought in new businesses such as logging, coal mining, and tourism, and the lodges, guiding, and outfitting support for hunters and sightseers. These successive changes – the national park, Dominion Forest Reserve, railways and resource and tourism businesses – all brought new people to the valley to staff their respective operations. However, these newcomers had been preceded by many generations of people whose families had been living in and around the Athabasca valley for centuries. The First Nations and Métis people had roamed to and through the country making seasonal use of the resources of the broader region, generally between the upper Fraser River and Lac Ste. Anne, and from Grande Cache and the Smoky River south to the Brazeau and North Saskatchewan rivers. They travelled variously on foot, by canoe or horse and with dogs, camping in tepees or cabins.

These people and their ancestors had experienced change and contributed to it for centuries before Jack Glen arrived. An historical perspective can help us see more clearly the importance of the forestry work in which Glen was involved and the context in which he made some of his observations.

Opposite page: Pack String crossing the Athabasca River on the old Canadian Northern Railway bridge, early 1940s.

GORDON WATT COLLECTION, ALBERTA FOREST HISTORY PHOTO COLLECTION

Forest Ranger and child on wagons with teamsters and oxen hauling freight for the Grand Trunk Pacific Railway on the Athabasca Forest, c. 1911

DOMINION FORESTRY BRANCH, LIBRARY AND ARCHIVES CANADA E008439149

Prehistory and Fur Trade

The earliest archaeological evidence of people in the area is of a camp near the Snake Indian River that was occupied about 10,000 ago. It was probably not a permanent settlement; people may have used the valley as a corridor for hunting or just passing from one valley to another. Travel in those days was slow, and a party might tarry for weeks, months or even years if the hunting, fishing and gathering opportunities were good enough to sustain them.

An archaeological dig at Patricia Lake, just north of Jasper, revealed obsidian flakes from Mount Edziza in the Coast Range, nearly a thousand kilometres to the west, brought to the Jasper area between 4,000 and 2,400 years ago. Other excavations at Patricia Lake suggest the site was also in use between 2,400 and 1,200 years ago. Long-distance trading and successive occupation were clearly evident.

Initially, the main influence in this region over the last 300 years seems to have been from the Plains people to the south. Then, in the early 1800s, the Beaver Indians dominated the foothills between the Athabasca and Peace rivers, a few Shuswaps lived in the Jasper and Mount Robson areas, and Stoneys, or Assiniboines, prevailed to the south. At this time the Cree began to increase their presence in the west and became strongly influential in this region. Characteristic of all these people was their ability to travel in response to seasonal availability of food and opportunities to trade with – and sometimes do battle with – adjacent people. Especially important were buffalo and moose on the eastern slopes and salmon west of the Rockies. The people also used fire to modify the landscape to support their hunting, gathering and travelling activities.

The stage for European influence in this area was set in 1670, the year the Hudson Bay watershed – including the entire Saskatchewan River basin – became part of Rupert's Land over which the Hudson's Bay Company (HBC) had exclusive rights to furs and trade. The adjacent lands to the northwest, including the great Athabasca–Peace–Mackenzie River basin, were also British, part of their North-Western Territories.

The fur trade was slow to make its mark in the upper Athabasca region. Initially the HBC built trading posts along the lower reaches of major rivers down which natives would travel with furs and return with trade goods. The first HBC trader in the area was Anthony Henday who wintered near present-day Edmonton in 1754–55. Then, in 1780 a group of Scots traders in Montreal founded the North West Company (NWC) to compete with the HBC. They explored and moved west and north, bypassing the HBC posts and taking trade to the Aboriginal people where they lived. Their rivalry was such that by 1795 both NWC and HBC had built posts at the mouth of the Sturgeon River, the first posts in the Edmonton area.

One of the competitive advantages of the NWC lay in their practice of hiring Iroquois (also referred to as Nipissings or Algonquins) on three-year contracts as canoemen and trappers. The Iroquois, an eastern tribe from the Quebec-Ontario area, were among the first to use steel traps, while Western tribes were still using deadfalls or digging beaver out of their lodges. After their contracts were up, many Iroquois stayed in the West as "freemen," trapping and trading independently, often taking wives from local tribes such as the Cree. Some Iroquois "freemen" appear to have been trading in the upper Athabasca region as early as the 1790s. In 1859, James Hector of the Palliser Expedition described camping in January with a group of free-trading Iroquois along the Athabasca River near the mouth of Oldman Creek. They had learned the Cree language and intermarried with the Cree of Lac Ste. Anne.

European traders often took Aboriginal "country wives," and their children became known as Métis or mixed-blood people. Along with other Aboriginal people, the Métis were important contributing members of the fur trade and brigades. Among these were descendants of Henry John (Harry) Moberly, HBC chief trader in Jasper from 1858 to 1861. Moberly married Suzanne Karakonti (Kwaragkwante[2]) and they had two sons, Ewan and John, who stayed on with Suzanne in the upper Athabasca Valley when Henry John left. They settled on homesteads in what is now Jasper National Park in the 1890s, along with two of Ewan's sons, William and Adolphus, and the families of Isadore Findlay and Adam Joachim. They continued the traditional practices of meadow burning in the spring to keep the valleys open for travel, to encourage grass and hay for their horses and to encourage range for bighorn sheep. Descendents of these families were among the Aboriginal and Métis people living in the Grande Cache area mentioned by Jack Glen.

David Thompson was the first European to visit the region. In the fall of 1810 his route to the Columbia River through Howse Pass was blocked by the Piegans (Stoneys) who were incensed that he had traded guns to the Shuswap, their traditional enemies to the west. Instead, Thompson led his brigade north from the Brazeau River to the Athabasca with his Iroquois guide Thomas who knew of a pass from the Athabasca. They camped for over three weeks somewhere around Brûlé Lake in December to hunt buffalo and make sleds and snowshoes, then crossed Athabasca Pass to the Columbia in January 1811. This established the course of the trans-Canada express brigade route between York Factory on Hudson Bay and the mouth of the Columbia River on the Pacific Ocean. The brigades ran through Athabasca Pass for over 40 years – Canada's first "trans-Canada highway."

[2] There have been many spellings of Suzanne's family name. Karakonti appears on her certificate of marriage. Some family members prefer Kwaragkwante, the question is being discussed.

National Parks and Forest Reserves

In 1867 four of the original eastern provinces entered Confederation to form the Dominion of Canada. Sir John A. Macdonald, first prime minister, undertook to acquire Rupert's Land and the British North-Western Territories. The deal was successfully concluded in 1870 and these additional lands (and their inhabitants) became Canada's North-West Territories. Alberta and Saskatchewan became provinces in 1905, but their forests and other natural resources remained under Dominion of Canada control until they were transferred to the provinces in 1930.

Given the competitiveness of Canada's southern neighbour, the United States of America, it was important to Macdonald to get British Columbia to join Confederation (1871) and to build a railway to connect eastern Canada with the Pacific Coast (1885).

By the late 1880s, the effects of uncontrolled land clearing and logging led to the start of conservation movements. In 1884 the Dominion Lands Act was amended to enable "the preservation of forest trees on the crests and slopes of the Rocky Mountains … for the proper maintenance throughout the year of the volume of water in the rivers and streams." This set the stage for the national parks and forest reserves that followed. The first area set aside was at Banff in 1885. What was to become the Dominion Forestry Branch was set up in 1899. The first Forest Reserves Act was passed in 1906, which included the Cypress Hills, Cooking Lake and Kootenay (Waterton Lakes) areas, and the Dominion Forestry Branch began surveys to identify suitable locations for other forest reserves, including the eastern slopes of the Rockies.

Jasper Forest Park was created in 1907. That led to the removal of the six homesteading Métis families in 1910, since the new administration would not permit settlement, burning or hunting within the park. John Moberly and his family moved to Prairie Creek near Hinton and the others moved to the Grande Cache area where they built homes, developed farms and continued to practice their traditional way of life within the constraints of the new order with its rules, regulations and officials.

The Rocky Mountains Forest Reserve (RMFR), which lies along the eastern slopes, was first defined in 1910 and new boundaries were extended in 1911 and 1913. The RMFR was divided into five "forests" for administrative purposes. From the south these were the Crowsnest, Bow River, Clearwater, Brazeau and Athabasca. The Athabasca Forest was the most remote and resources scarce, so was the least developed in 1920. In later years the Athabasca Forest was often combined with the Brazeau under one supervisor at Coalspur as a cost-saving measure, but the ranger station at Entrance remained the central post for the Athabasca.

Before 1930, the federal Department of the Interior was responsible for national parks and forest reserves, so boundary changes between them were common. Jasper Park was reduced in 1911 and expanded again in 1914 and 1923. Final boundary adjustments to the parks and forest reserves were made in 1929 and 1930 in preparation for the Transfer of Resources to Alberta. Lands of high park value in the forest reserves were transferred to parks in exchange for land of high commercial value, such as coal mines, in parks. In 1930 the Alberta Forest

Service (AFS) took over administration of the forest reserves, including the Athabasca Forest. Jack Glen had started in 1920 while it was a Dominion Forest; he and most of the staff stayed on after the transfer to the AFS[3].

[3] P.J. Murphy et al. (2007). *A Hard Road to Travel.* Hinton: Foothills Model Forest and Forest History Society.

Entrance

The headquarters of the Athabasca Forest was logically located at Entrance, from which the pack-trail systems led north and west. Entrance was the last station on the CNoR before the Jasper Park boundary in 1914, so the name "Entrance" symbolized a gateway to Jasper as well as to the mountains and the Yellowhead Pass. In addition to Forestry staff, residents of Entrance included the railway station agent and track maintenance workers, the Woodley family with their store and restaurant, guides and outfitters, trappers and others. There was a low-water ford across the Athabasca River at Prairie Creek and CNoR built a ferry there around 1911. The first surveyed road in the area led from the ferry to Entrance in 1917.

The railway grade west of Entrance ran along the base of the high steep bank of the Athabasca River. It had been hastily built by CNoR and needed so much maintenance that in 1926 the CNR built a new bridge about three kilometres farther upriver toward Brûlé Lake and used a combination of the old GTP grade and new construction for about five kilometres to the new bridge. Entrance station was abandoned and the former unused GTPR station at Dyke was refurbished and renamed Entrance. This became the "new" Entrance and the original community became known as "Old Entrance". Tom Monaghan bought the Woodley business in 1925, built a new store and post office at the new Entrance in 1927 and closed the old one, but the Forestry headquarters and most of the independent residents stayed on. Later, planks were laid across the abandoned railway bridge, making it possible to cross the river at any time, by foot, horse and motor vehicle.

This was the setting into which Jack Glen moved in 1920. Entrance was a remote and pioneering community; almost every resident had come from somewhere else, bringing their own values and insights and determined to make this new area theirs. Jack Glen did his part to establish a "Forestry" presence. As his son Jim later described, he took his job seriously, did his work well and was influential in developing trails, cabins and lookouts to better protect the Athabasca Forest from fires and its wildlife from poachers. He certainly seems to have enjoyed what he was doing during his twenty-five years with "Forestry" before moving to his new home on the Skeena River in British Columbia.

I had contemplated staying with the Mounted Police until I retired, but I noted that while I was enjoying myself, I was obviously getting nowhere and had hoped to be sent north to the Arctic where Mounties had a more meaningful role as a police force.

Jack Glen (l) with a group of fellow Royal North West Mounted Police (RNWMP) officers, late 1910s.

JACK GLEN PHOTOS, ALBERTA FOREST HISTORY PHOTOGRAPHIC COLLECTION

CHAPTER ONE

FIRST DAYS ON THE TRAIL

From Mountie to Ranger: A New Career in the Athabasca Forest Reserve

Late in April of the year 1920 my chum and I were sitting in our room in the Royal North-West Mounted Police barracks in Edmonton, Alberta, we were known at that time as "G" Division. It was a Saturday morning. We had been to stables, had breakfast, made our beds, tidied our rooms and were awaiting room inspection, the custom on Saturday mornings. We had been discussing our future plans as we both had submitted our resignations, effective on May 1.

Inspector Montezambert and the orderly sergeant entered the room. We stood at attention until the inspection had duly taken place, whereupon the inspector dismissed the sergeant and told us he wished to have a chat with us. Offering cigarettes, he asked if there was any discontent with our treatment by the Force behind our resignations. "I understand," he continued, "that many of our recruits get discontented because promotion comes so slow in the Force." "No, sir," I replied. I explained that I had contemplated staying with the Mounted Police until I retired, but Alberta at that time had its own provincial police force, and Mounties in the province were relegated to revenue work, naturalization duties and picking up military evaders. I noted that while I was enjoying myself, I was obviously getting nowhere and had hoped to be sent north to the Arctic where Mounties had a more meaningful role as a police force. The inspector was silent for a time, then suggested we might reconsider our decisions and that in my case an Arctic posting might come sooner than I thought. (As it transpired, in a little over a month most of the men from "G" Division were posted to the north.)

But I had made my mind up to quit, and told him so. My partner, Eddie, was also determined to quit, albeit for a different reason. He was about to marry, did not consider a constable's salary sufficient to support a wife, and had an offer of a good job with Alberta Government Telephones.

The inspector then asked my plans for the future. I told him I had made application to get on the Dominion Forest Service. The inspector approved of this move and said there was no other job outside the Mounted Police that he would sooner have himself. "Your Mountie training will be most beneficial to you in your new environment," he added.

So I left the Force, with little money saved and no word on whether my job application was accepted by Forestry. Work was plentiful for skilled labour, but I could not take permanent work while awaiting word on a Forestry job, so I went to the stockyards and made a few dollars doing odd jobs.

I considered farming as a last resort. As a returning First World War veteran I could have a Soldier Settlement Farm just for the asking, and I made a visit to the Soldier Settlement Board headquarters to that end. The examining officer proved to be the arrogant type who treated me as though I were still in the army

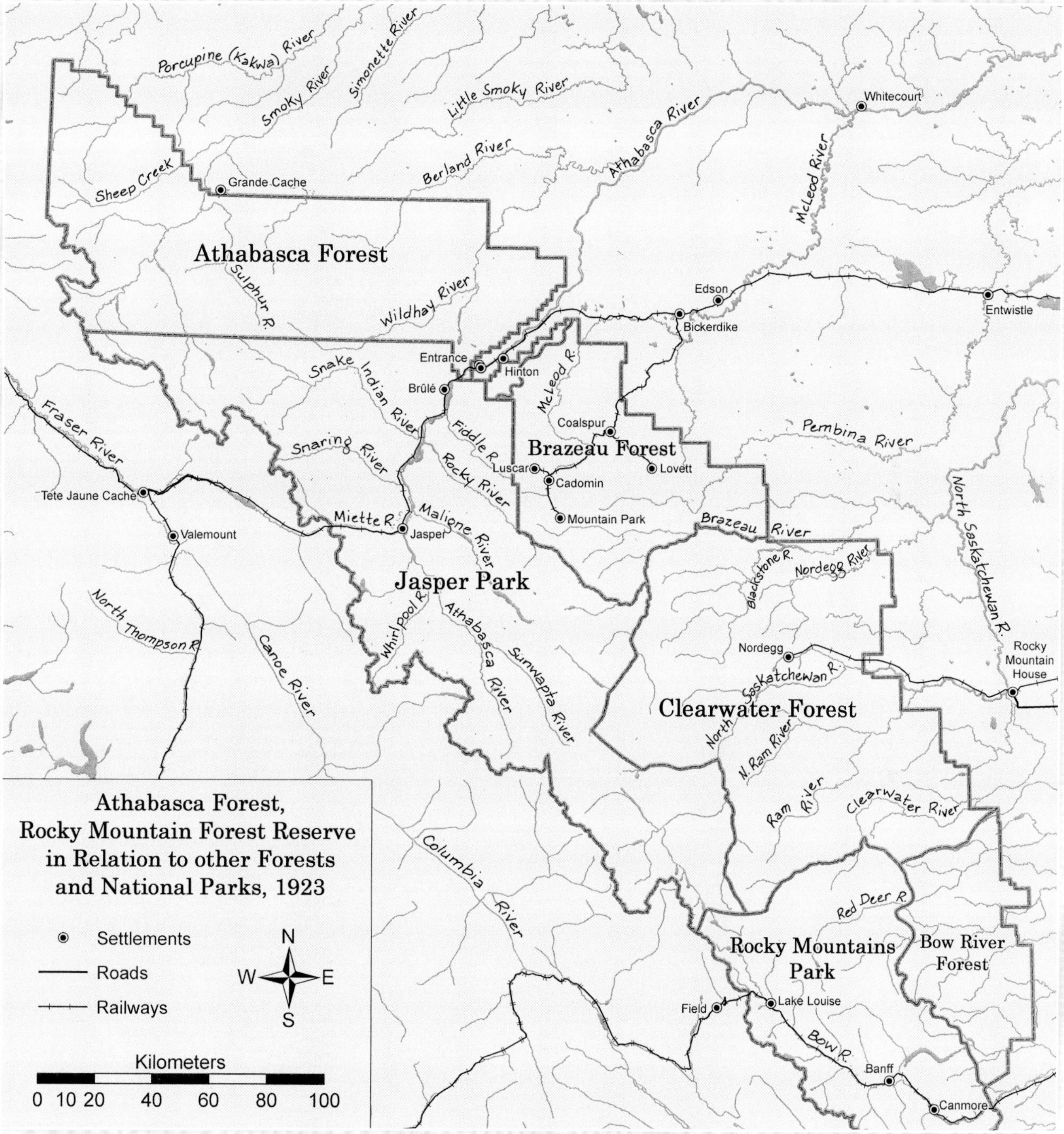

Map 1. The Athabasca Forest in relation to other National Forests and Park Reserves, 1923.

Dominion Forestry Branch ranger's badge.

BOB STEVENSON, ALBERTA FOREST HISTORY PHOTOGRAPHIC COLLECTION

Early Alberta Forest Ranger, Clearwater Forest 1930s. Early Rangers were not issued uniforms, only badges denoting their status in the Dominion Forest Service organization. Some rangers chose to carry handguns but in Canada permits were required to do so after 1913.

ALBERTA FOREST HISTORY PHOTOGRAPHIC COLLECTION

and would have to dance to his music. I would rather dig ditches than submit to that kind of kowtowing, and I terminated the interview abruptly before I blew my top.

I had been boarding with my married sister and brother-in-law in Edmonton, and when I returned that evening I was told to call a Mr. Badgley at the King Edward Hotel. Wondering who this might be, I called the number given and all doubt was removed when he answered, "Badgley from the Dominion Forest Service here. Are you the gentleman who applied for that civil service position in the Athabasca Forest Reserve quite some time ago, and if so, are you still available?" I answered yes to both questions, and we agreed to meet the next morning. Wild horses could not have kept me away.

Next day, we shook hands and he sized me up for a few seconds before getting down to business. Mr. Badgley was a tall, heavy-set man I judged to be in his sixties, and I was impressed by him from the start. He explained he would have contacted me sooner, but had been ill, and besides with the late spring and lack of grass it had been expensive to feed the Forestry horses, leaving little money for new recruits. He outlined the job and made no bones about its difficulties and hazards.

"It's a dangerous job," he said, "and you'll have to be alert at all times. It's one of the best big game countries in the North American continent. Especially now, after the war, due to lack of ammunition the game animals have been left unmolested." He continued with a twinkle in his eye, "But that doesn't make the grizzly bear any more kindly disposed towards the human race. There'll be swollen rivers to cross, snow slides in the mountains, dangerous ice, venomous insects and forest fires. On the credit side of the ledger, our Forest Reserve has everything that a young red-blooded man could wish for. The fishing is wonderful and big game is plentiful. Tourists from every part of the world are spending thousands of dollars to see what you will be seeing free. You may run into a few lawless characters, but with your Mounted Police training you would no doubt know how to handle them."

Needless to say, my answer was yes. Mr. Badgley then proceeded to outline my duties. He told me that at that time there were only two main trails in the Athabasca Forest Reserve. There were a few miles of wagon road also, and the Service also owned a railroad "speeder," but it was seldom used because he considered it much too dangerous. Several new branch trails were to be cut, not only for forest protection but also to make the country more accessible to tourists and big game hunting parties. I was to go out on the Mountain Trail, travel as much as I liked and get acquainted with the country. Then I was to be put in charge of a trail crew and build the proposed trails. I was to provide myself with a bedroll and rifle, horses, and riding and pack saddles, this being the sole method of travel in the Forest Reserve. He said I could dress as I pleased, and I would be issued with some binoculars and with a badge of office that must be worn conspicuously.

He advised me to buy my horses in Edmonton, since the previous winter had been severe. Many horses and cattle had died, and consequently the prices being asked in the Forest Reserve were prohibitive. He then wished me good luck and told me to get out to Entrance as soon as possible.

My New Frontier Home

I lost no time in making my purchases and arranging for transportation. I bought two horses at the stockyards and travelled with them in the caboose. I was on my way. Mr. Badgley and the ranger staff met me at Entrance, I was assigned a bunk in the ranger cabin and told I could suit myself about batching or board at the restaurant until I left for my district.

Entrance. What a sleepy little place, I thought, and wondered if anything ever happened to break the monotony. Little did I know that during the many years that I would live around that same sleepy little place I would see enough excitement, tragedy and drama to last the average man a lifetime.

Entrance at that time was classed as a hamlet. It consisted of a general store, restaurant and schoolhouse. The post office was in the store and the poolroom in the restaurant. A few trappers' shacks, owned by the Woodley brothers – Earl and Roy – who ran the store, and the Canadian National Railway with its water tank completed the hamlet. The Forestry buildings, all of log construction and very substantial and cosy, consisted of the supervisor's residence, ranger cabin, barn

Ward Badgley, c. 1890s.

ROY WOODLEY COLLECTION, COURTESY OF MARY LUGER

William Ward Badgley

William Ward Badgley (1867–1929) was born in Ontario and married Sarah Stiles (1865–1958) in Rawdon Twp, Ontario in 1890. Sarah had graduated from the Ottawa Normal School with a teacher's certificate in 1888. Ward and Sarah farmed near Barrie, Ontario after they were married, then they operated a hardware store and Massey Sawyer steam engine dealership in Yorkton, Saskatchewan for several years.

They came to Alberta in April 1914 when Ward was hired as forest ranger at Sawridge, near Slave Lake. He transferred to Entrance in August 1917 as supervisor of the Athabasca Forest for five years to 1922.

Ward and Sarah had four children: John, Alberta, Florence and Edna. Earl Woodley met Edna in Entrance and they were married in Victoria, B.C. in 1923. They moved to Red Pass, B.C. to develop their store, mail order and hotel businesses while Earl's brother Roy ran the business in Entrance.

Courtesy of Mary Luger

Entrance as seen from the south side of the Athabasca River, (r-l): Dominion Forestry Branch bunkhouse, ranger station and barns; Woodley's barn, store, restaurant, water tower, station buildings. Ted Hammer's house is up the hill on right side, later converted to the schoolhouse; Woodley homes are also up the hill on left side, c. late 1930s.

MARK TRUXLER COLLECTION, COURTESY OF JACKIE HANINGTON

and cache and were located on government property nearby. A beautiful fenced lawn and large shade trees surrounding the supervisor's house was an impressive sight, often admired and photographed by people passing on the train.

That first night, the place looked practically deserted till suppertime. I was the first to sit down at the counter and then the customers began to arrive and in no time the counter stools were all occupied. They must have fed thirty or forty people that evening. The pool table in the meantime was doing a land office business and it was the same in the store. Evidently closing time did not mean a thing to those people. Some were buying while others were sitting in Nail Keg Alley, so called because Mr. Woodley had placed a plank on top of a number of nail kegs for the convenience and comfort of his patrons. His place was thick with tobacco smoke.

My attention was drawn to a tall, powerful-looking man of about sixty who evidently had the floor, but from where I was sitting I could hear little of what he was saying. I asked the man sitting next to me who he was.

"That," he said, "is Montana Pete[4]." Years later Howard O'Hagan mentioned him in "Montana Pete Goes Courting" in his book *Wilderness Men*. I learned his history later on and he must have led a colourful life. He was raised in the Mormon faith in Utah and emigrated to Canada at an early age. He lived with the Blackfoot and Cree Indians and learned their language. For a number of years he acted as interpreter for the North-West Mounted Police at Battleford. He freighted

Montana Pete

Jedidiah (Jefferson) Leatham was born in Utah in 1865, the child of a polygamous marriage. He ran away from home at age 12, eventually arriving at North Battleford the year after the Riel Rebellion. He joined the Yukon Gold Rush in 1898. While there, he drove mail by dog team for the North-West Mounted Police, and narrowly missed being caught in the 1898 Chilkoot Pass Slide. In the early 1900s, he worked as a packer on the old Grand Trunk Pacific rail line from Edmonton to Tête Jaune Cache. Eventually, he homesteaded just north of Old Entrance and ran a trapline up the Wildhay River above Solomon Creek. When he turned 65, he had difficulty proving his age until he was able to produce a transcript from an old family Bible, and was awarded a pension of $15 per month (1935). Eventually, he sold his homestead to Harry Davison, who owned the nearby Entrance Ranch, and settled in Hinton where he remained the rest of his days. Pete's Flats, up Solomon Creek towards Rock Lake, was named after him. (Source: Hazel Hart, *History of Hinton*, Friesen, Edmonton, 1980).

Montana Pete, 1927.

ROY WOODLEY COLLECTION, COURTESY MARY LUGAR

for the contractors building the Canadian National Railway at Hinton and Entrance. His real name, I believe, was Jefferson Leatham, and he had a homestead not far from Entrance.

That night, too, I inquired about Entrance and how it had been named. I was told that it was the entrance to the mountains and the Yellowhead Pass and I thought the name very appropriate. The next day was Sunday so I saddled up my horse and rode towards the mining town of Brûlé. I had been told to look out for the Black Cat, or Brûlé Cat as it was sometimes called. It was formed by green timber growing up after a fire, and it certainly bore a strong resemblance to a cat with its back arched, and in winter the snow appeared to deck it out with a white ribbon round its neck. In later years, a new forest fire burned the legs and now the effect is largely gone.

The next day everybody drifted over to the depot to watch the Silk Train go through. This was a special express train with armed guards that had the right of way from the Pacific to the Atlantic. Its cargo of raw silk was highly valuable and very perishable, so no time was lost at divisional points. A new crew and locomotive would be there waiting to take over and the guards ate and slept in a special car next to the caboose. It passed through the depot at terrific speed, almost a blur, and was a great sight to watch. Strangely enough I never heard of a wreck with any of those trains.

That afternoon there was more excitement. An inquest was held on the death of an Indian who had been hit by a train near the Athabasca Bridge. It seemed there had been some drinking going on and foul play was suspected. One of the suspects was a big man, very dark and swarthy. But when he emerged after a severe grilling by a Provincial Police officer, his complexion had changed to a sort of sickly yellow. And so it went; truly the "Old West" still lived.

My conception of Entrance as a sleepy little place had undergone a change and I felt as if I was watching a thrilling Western drama unfold before me. We had no saloons, but I will admit that a bootlegger did a thriving business, and a gentleman from Georgia distilled a potent brew of White Mule, or "Cawn Likker" as he called it. Poker games were played quite often in the shade of the water tank, with the sky the limit. Fist fights were common, but nothing in the line of shooting more serious than taking pot shots at the metal ball that acted as a water gauge atop the water tank, or shooting off the odd glass insulator on the telegraph line.

The area had its share of crime also. Alberta had Prohibition at this time and Saskatchewan was wet. A box-car load of liquor – sealed for the trip – was being shipped to British Columbia. Inspection at Edson, a divisional point, showed the seals were unbroken, but on arrival at Jasper the seals were found to be broken and two kegs of rum had been stolen. That presented the revenue officers with a mystery that was never solved. The freight train had made only one stop at

Opposite page: The Black Cat stand near Brule. This patch of older spruce strongly resembled a black cat. The spruce originated from a forest fire around 1808. In 1896, another fire burned through the area, leaving this cat-shaped patch of trees untouched, c. 1920s.

JACK GLEN PHOTOS, ALBERTA FOREST HISTORY PHOTOGRAPHIC COLLECTION

Old Entrance, 1916. (r-l) Stables, Woodley Store, Restaurant, water tower with steel ball on top. The Woodley residences and schoolhouse are on the hill behind the store.

GLENBOW ALBERTA MUSEUM AND ARCHIVES NA-379-5

Opposite page: Map 2. Dominion Forestry Branch patrol trails and cabins, 1920. The two main trails in the Athabasca Forest were the Mountain Trail and the Lower Trail.

FOOTHILLS RESEARCH INSTITUTE HISTORY SERIES

Entrance where it took on water. One of the many theories advanced was that a gang of probably two or three men had broken the seals, opened the car doors on the blind side of the track and entered the car. That part sounded reasonable as the train was travelling at night, but the rum could not have been unloaded there and carried off without being noticed, as at the time trucks and automobiles were unknown in that part of the country.

A few miles west of Entrance the railroad track crossed a shallow bay on Brûlé Lake where the kegs might have been rolled off in the water and the men could have jumped after them. Tracks of horse and man were noticed along the shore which gave rise to that theory, but all signs vanished in the dense timber nearby.

Soon after I arrived, the safe at the Blue Diamond Mining Company's office in Brûlé was blown open and the payroll stolen.

First Patrol into the Athabasca Forest Reserve

One day soon after all this excitement three of us started out up the trail to the first ranger station, called Winter Creek. This was where the pack and saddle horses belonging to the Forest Service were wintered. Luckily we had a fenced pasture to turn our horses into and hay to feed them, as I had been warned that horses brought in from outside were liable to cause trouble, and I found that mine were no exception to the rule.

Spring was slow in coming to the Athabasca Forest Reserve in 1920, the trail was muddy and the grass was just starting to show green. Ranger Charlie Milendorf[4] had a notion of turning back but thought if we could make it for two more days we would reach good horse feed along the Berland River where bunch grass was plentiful. We reached the Moberly Creek ranger station next evening, but unfortunately there was no pasture there, so we turned our horses up the creek and hoped for the best. We were now on the Lower Trail, having passed the junction with the Mountain Trail about three miles back, as no horse feed was

[4] Charles F. Milendorf worked four years for the Dominion Forestry Branch as a forest ranger on the Athabasca Forest from 1918 to 1921.

Riders and ranger at the Moberly Ranger Station, c. 1920s. A note identifies one of them as Louie Holm.

JACK GLEN PHOTOS, ALBERTA FOREST HISTORY PHOTOGRAPHIC COLLECTION

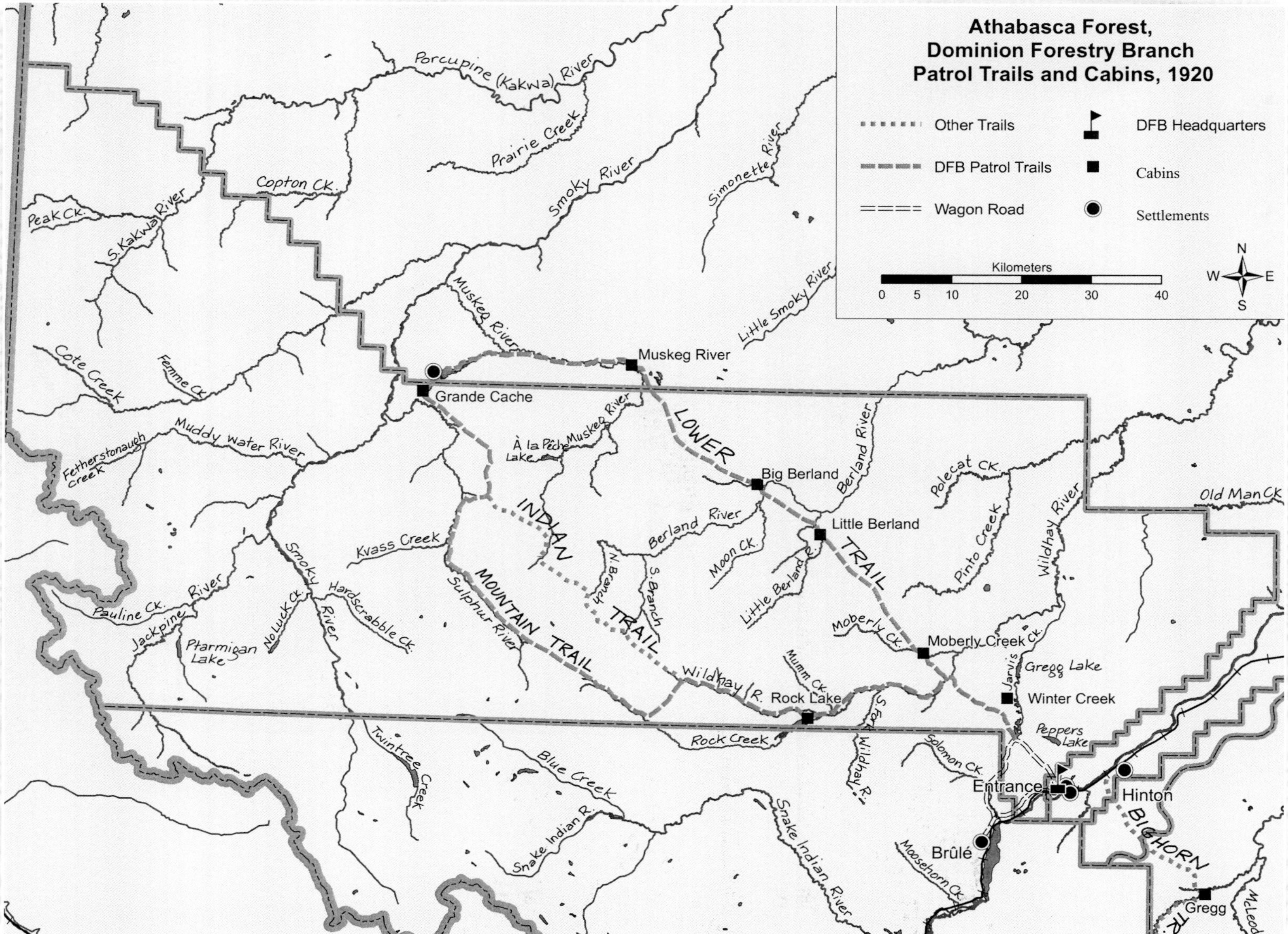
Athabasca Forest,
Dominion Forestry Branch
Patrol Trails and Cabins, 1920
Other Trails
DFB Patrol Trails
Wagon Road
DFB Headquarters
Cabins
Settlements
Kilometers
0 5 10 20 30 40
N
W
E
S
Porcupine (Kakwa) River
Prairie Creek
Smoky River
Simonette River
Copton Ck.
Peak Ck.
S. Kakwa River
Little Smoky River
Muskeg River
Cote Creek
Femme Ck.
Muskeg River
Grande Cache
Muddy Water River
Fetherstonaugh Creek
À la Pêche Lake
Muskeg River
LOWER
Big Berland
Berland River
Polecat Ck.
Old Man Ck.
Little Berland
INDIAN
Berland River
Pinto Creek
Wildhay River
Kvass Creek
Moon Ck.
TRAIL
N. Branch
S. Branch
Little Berland
Smoky
River
Hardscrabble Ck.
Sulphur River
MOUNTAIN TRAIL
TRAIL
Pauline Ck.
River
Jackpine
Ptarmigan Lake
No Luck Ck.
Moberly Ck.
Moberly Creek Ck.
Jarvis
Gregg Lake
Winter Creek
Wildhay R.
Mumm Ck.
Rock Lake
S. Fork Wildhay R.
Rock Creek
Peppers Lake
Solomon Ck.
Twintree Creek
Blue Creek
Entrance
Hinton
Snake Indian R.
Snake Indian River
Brûlé
Moosehorn Ck.
BIGHORN
TR.
Gregg
McLeod R.

Ward Badgley (r) crossing small stream, around 1916.

ROY WOODLEY COLLECTION, COURTESY OF MARY LUGER

available there. Ranger Louie Holm was to travel with me as far as Rock Lake Cabin, the first cabin on the Mountain Trail. It was decided that he should leave me there and return to Entrance, but the best laid plans of mice and men will often "gang agley" (go awry), according to Robbie Burns, and the lowly cayuse[5] can shape the destiny of a man in no uncertain manner.

Daylight was just breaking when I was awakened by the sound of horse bells. I partly dressed and rushed outside. A mare and her colt were galloping past the cabin closely followed by two horses. I might have saved the day had I picked up a bridle but the thought uppermost in my mind was to try to get ahead of my horses. I followed them almost to the Wildhay River and did get my hands on my saddle horse, but unfortunately all I had to put around his neck was a piece of string. I petted him and told him what a fine little horse he was, but he did not seem to be paying much attention to me and as his mate was still tagging along with the others, and he commenced to rear and plunge. The string broke and he lost no time in joining his mate and the mare and colt. I knew it was hopeless and went back to the cabin.

The boys had breakfast ready and they had telephoned back to Winter Creek to get our teamster to try and head off the culprits. I telephoned Mr. Badgley and told him the story. He laughed and said I had now gone through my initiation and was now a member of the Loyal Order of Dismounted Forest Rangers. The teamster had no luck stopping the horses, so Louie Holm and I went back to Entrance to wait until the horse feed improved. We busied ourselves for the next ten days or so blasting stumps and clearing a site for a new ranger cabin. My horses were finally located and brought back. Luckily they ran into an old burn where the windfall was piled so high that only a bird could have crossed it. The homing instinct must have been strong in them as they were headed east towards Calgary, their home range.

Louie Holm went back with me to Rock Lake but we had no more trouble with the horses as I had bought two pairs of hobbles, which limited their activities considerably. Louie stayed with me for two or three days and helped me with the firewood, as the pile that should have been there was almost non-existent. We fished for rainbow trout in the evenings.

I was sorry to see Louie go but I was too busy exploring to feel lonesome and besides I had to get used to it.[6] There was no telephone line on the Mountain Trail as the building of it would have presented quite a problem and the cost of maintenance would have been prohibitive. It would be many years before we had radio communication. I was anxious to see the rest of the district, but Mr. Badgley had told me I had better stay at Rock Lake for a time. He was trying to hire an assistant for me. If he did, the two of us could move up to Eagles Nest Pass and finish building a cabin there that had been started before the war.

Rock Lake was a lovely spot. You could stand in the cabin door and with the aid of binoculars pick out elk and sometimes mountain goat on the mountain

[5] Glen was probably using "cayuse" in the informal sense that many cowboys used when referring to their horses. However, the Cayuse Indian Pony is actually a distinct breed of horse that originated in the 1800s. Small and stocky, it has high withers and an unusually long canon bone (the main bone from the front knee to the ankle joint), which gives the body a distinctively sloped appearance. Source: Breeds of Livestock, Oklahoma State University, Department of Animal Science, updated 1996. www.ansi.okstate.edu/breeds/horses/cayuseindian/.

[6] Louis Holm joined the Dominion Forestry Branch in 1918 as a forest ranger and served on the Athabasca Forest until the end of the season in 1929 living in Entrance. (Feddema-Leonard 2007)

across the lake. Canada geese with their honking – always a welcome sound to my ears – heralded the coming of summer and another mating season. Our horses craved salt, which we used to feed them at the hitch rail near the cabin. The deer got to know about this too, and sometimes in the evenings when I was reading I would hear a sound outside. Tiptoeing to the window I could watch those lovely creatures searching for the odd morsel of salt that the horses might have overlooked. They appeared so dainty and graceful in their movements that to shoot one of them would have seemed like committing murder.

Yes, I grew very fond of Rock Lake and in the years that followed I almost regarded it as home. Nellie McClung, better known at the time as Janey Canuck, had visited Rock Lake before me, leaving evidence of the buoyant mood inspired by the area in verses of Psalms written on the cabin walls. "I to the hills will lift my eyes," was part of one verse that I thought particularly suitable for the occasion.

Rock Lake Cabin and Corral, c. 1920s.

MARK TRUXLER COLLECTION, COURTESY OF JACKIE HANINGTON

One day I took an axe and started up the trail to cut out the windfall and reached the Wildhay River, which we got into the habit of referring to as the Hay. Ordinarily, we took our saddle horses to ford the rivers, but this day I was on foot. I looked for a growing tree that I could fell across the river and get over dry shod, but I could not find one and decided to go further upstream. Rather than make a lengthy detour, I decided to fight my way through a dense jungle of willow and alder.

Suddenly I heard a scratching sound and caught a glimpse of something black. A bear cub perhaps! And I could possibly be standing between the cub and its mother. Now I have always had a great respect for Mother Bruin, so I turned back in the direction from which I had come and did not lose any time either. I got back to the open gravel bar where I started from and sat down and wiped the sweat from my forehead. I watched that clump of willows closely, quite prepared

Meadow at the Big Bend of the Snake Indian River, Athabasca Forest, Supervisor Stan Clark on patrol, 1915.

DOMINION FORESTRY BRANCH, ALBERTA FOREST HISTORY PHOTOGRAPHIC COLLECTION

to continue my flight should circumstances warrant so doing. Then my bear cub emerged in the form of a large black porcupine. Feeling a great deal better, I ate my lunch and had a smoke. When I went back upriver I found a suitable tree, which I felled and got across again without further incident. I kept on up the trail chopping out old windfall and admiring the scenery until I figured it was time to go back.

Then I made an odd discovery. A beautiful grassy hill caught my eye. As I got closer I thought what a wonderful spot for horses to graze and I noticed that it was terraced. It looked almost as if it had been the work of man. The terraces or paths were spaced about six feet apart from the bottom to the top of the hill and followed its contour perfectly. I could not understand it so the next time I met Ranger Milendorf, I described the spot to him and asked if he could explain the mystery.

He recognized the place at once, and told me that the buffalo were responsible. According to the oldest Indians, this was once wonderful buffalo country and when they grazed they followed the contour of the hill and cropped the vegetation above them. Then when they reached the end of the grass they stepped up about six feet and reversed the procedure. They certainly used their heads.

I asked him what happened to the buffalo and he told me that a hard winter and wolves finished them. Then I remembered having noticed broad skulls in the vicinity, with the inner core of horns still attached. The outer horns would soon have been chewed up by porcupines, along with the rest of the skeletons.

Nellie McClung

Nellie McClung (1873–1951) was a teacher, author, suffragist, social reformer, lecturer and legislator. She was one of the "Famous Five" (with Irene Parlby, Henrietta Muir Edwards, Emily Murphy, and Louise McKinney) who, in 1927, submitted a petition for an interpretation of the word "Person" in section 24 of the *British North America Act, 1867*. The decision on 18 October 1929 (the Persons Case) found that "person" includes female persons, making women eligible for appointment to the Senate of Canada. In 2000 the Canadian government erected the "Women Are Persons" statue on Parliament Hill, an honour previously reserved for prime ministers and royalty.

[Ed:] Mr. Glen erroneously refers to her as "Janey Canuck," but it was her colleague Emily Murphy who actually wrote under the name "Janey Canuck."

Sources: *The Canadian Encyclopedia*, 2nd ed. Hurtig, Publishers, Edmonton, 1988; and *The Canadian Encyclopedia*, online edition, 2001. http://thecanadianencyclopedia.com/ (under Emily Murphy).

A New Assistant, Building the Eagles Nest Cabin

One evening at Rock Lake I was standing by the shore watching a kingfisher catching his supper when I thought I heard a voice or voices down the trail. Nor was I mistaken for in a short time a tall, gaunt man appeared leading a saddle horse and two pack horses. He had what looked like a violin case under his arm and he was still talking, though there was obviously no one else about. I concluded he was just one more soul who had missed too many boats.

Suddenly he looked up and saw me and asked if I was the District Forest Ranger. I said I was and asked if he would care to stay for the night. He thanked me and said he would. He laid the violin case, for such it was, very carefully against the cabin wall and set about unpacking. I busied myself getting supper for him. I had already eaten and when he had finished unpacking and had turned his horses loose, he rummaged through one of his packs and handed me some letters. One was from Mr. Badgley and I opened it first:

> Dear Glen,
> This will introduce you to the bearer, Samuel Munson,[7] whom I am sending you as an assistant. Sorry I couldn't find a man more of your own type but they just can't be got. The survey parties are snapping them up at a more attractive salary than we of the Forest Service are allowed to pay. Munson is a little queer but I am sure quite harmless, so if you can get along with him for the balance of the season I will try and find another assistant. Munson is a Trapper so should be quite handy in the bush.

[7] Sam Munson does not appear on any Forestry Branch staff list, so he was probably a part-time employee.

After I read the rest of my mail I sized Sam up. He was a man of about sixty and walked like a person who had had military training, but I guess I was mistaken, as he never mentioned the army to me. I tried to draw him into conversation, but apart from telling me that he always trapped on the Simonette River, and some of the most recent happenings at Entrance, he seemed rather reserved, so I assumed he was tired and did not press him for more.

He was Norwegian and had a very pronounced accent. "Yim" and "Yeff" were the two packhorses and his saddle horse went by the name of The Colt, but I'm sure that said horse would have been eligible to vote if horses had had the franchise.

Now that Mr. Badgley had provided me with an assistant, I set out with Sam's help to rebuild the cabin at Eagles Nest Pass. So far, Rock Lake was the only finished cabin on the Mountain Trail and a second one would add greatly to the comfort of the work.

We travelled to the foot of Eagles Nest Pass and I suggested that we make a comfortable camp before commencing the building. Sam thought this was a very good idea, so we built a wall of small logs about three feet high and chinked the

seams with moss. Then we made a frame of light poles on top of the logs and stretched our tent on that. A doorway was made in one end and at night we covered it with one of our canvas pack covers. A folding cook stove had been left by the men who started the cabin, so we dug a hole to set it in. If the weather was stormy we could eat and cook inside and when it was nice we ate outside. All in all, we were very comfortable.

The only fly in the ointment was that our horses wanted to go back to Rock Lake. As I mentioned before, I had hobbles, but if they were used too long the horses' legs became chafed and sore, so something had to be done about it.

I had hiked up into the pass and found that at one spot it was quite narrow. Here I asked Munson to build a drift fence, while I returned to Entrance to hand in my diary and buy supplies. A drift fence is made by falling trees one upon another until you have an insurmountable barrier, then a set of slip rails on the trail completes the fence. In later years, big game outfitters whose horses were affected with the wanderlust made good use of this fence.

The weather had turned extremely hot and the snow was fast disappearing from the mountains, causing high and muddy water. I had not given this hazard much thought and it was only when I called Mr. Badgley on the telephone that I realized the gravity of the situation. He told me that all the rivers were rising rapidly, the Athabasca was bank full, and he suggested that I make my trip to Entrance early in the morning, get the supplies and camp that night on the same side of the Wildhay. The water is always lowest in the morning. When I came to recross the river the following day, he said he would get Louie Holm to come down to the river. He was doing some telephone work nearby.

I followed the suggestions and everything went off according to schedule. When I was ready to cross again, there was Louie Holm pointing here and there and trying to tell me something, but the river was making such a noise that I could not hear a word that he said.

First I made sure that my pack was secure, then I tightened the cinches on my riding saddle and headed across. I noticed that the river was much higher than when I last crossed it, but still I was not worried. The same could not be said for Louie as he was running up and down on the other side and wringing his hands, plainly agitated.

I had a few anxious moments when my horses stopped to drink. I learned a lesson then that I never forgot. I looked down at the water swirling all around me and I felt so dizzy that I almost fell out of the saddle. Then I focused my eyes on a point high on the opposite bank and I felt all right again and crossed without further incident.

Louie was as pale as a ghost. "I thought you'd be drowned," he said. The sweat was dripping off him. I asked him what he had been trying to tell me.

"If your horse gets into difficulties," he said, "get out of the saddle over his rump, grab his tail and leave the rest to him. And I was trying to point out the

Opposite page: Pack train below Cathedral Mountain, Eagles Nest Pass, 1940s.

MARK TRUXLER COLLECTION, COURTESY OF JACKIE HANINGTON

spot where you should begin your crossing and where you should come out on this side. You should be thankful that you have good stout horses, otherwise I don't think you could have made it."

Next time I crossed the river I studied the ford carefully and noticed that a gravel bar ran diagonally across the river. The water was clear then, whereas it had been muddy when it was on the rampage, but I made a mental note of it for future reference.

The weather stayed hot, so hot in fact that Sam and I were forced to work in the early morning and in the evening. During the heat of the day, we would seek the shade of the timber and I would try to sleep. Pretty hard, though, for if Sam was not talking to me he was talking to himself.

Sam Munson's Story

As we got better acquainted Sam became more communicative, and by and by I learned his life history.

In his youth he had studied music and had made a career of it. He travelled practically around the world with an opera company. He not only played the violin but composed music as well. Every minute that he could spare he devoted to his work. Then the inevitable happened and he had a severe breakdown. The doctor told him to forget about music, get out in the country and relax.

Sam had been raised on a farm so he thought that besides recuperating he could make a comfortable living at farming, and he bought a mixed farm somewhere in Montana. It turned out to be anything but profitable and two years of drought put him entirely out of business, and he walked off the farm with only the clothes he was wearing, his cherished violin and very little money in his pockets.

His creditors kept hounding him wherever he went. He paid whatever he could until he was satisfied that he was out of debt, but they still kept after him and he finally decided to give them the slip. He crossed the border into Canada and in the end landed at Entrance where he managed to get a trapline out on the Simonette and Little Smoky rivers, where he hid out for several years, coming into town only once a year to sell his fur and get a grub supply.

The lawyers finally traced him and he showed me the letter they'd sent and asked me what he should do about it. I had come to the conclusion that the illness and quite possibly the financial losses had affected his brain, and the lawyers, noticing this, were just playing him for a sucker. I advised him to ignore the letter because by this time the case would be outlawed and they could hope to gain nothing through a lawsuit.

"By yimminy," he said, "that's yust what I vill do."

Sam read a great deal and had brought along a huge sack of adventure magazines. He believed every word he read. When he read stories about buried treasurer he would clip out the map showing the location of such treasure, until he had quite a sheaf of maps and charts, which he referred to as his parchments.

One day he had been gazing long and earnestly at one of these parchments and, turning to me, he asked if I could handle a machine gun. I said I could. Then he outlined a plan whereby the two of us could journey to a spot somewhere in the Pacific, dig up the treasure, and live in luxury the rest of our lives. He said he made quite a bit of money out of his trapline and could buy and equip a 28-foot cabin cruiser to take us to the treasure trove. My part of the scheme was merely to shoot any of the other pirates (or monkeys as he called them), should they try to interfere. "Now what do you think of that scheme?" he asked. I agreed that it was an easier way of amassing a fortune than working for the Dominion Government as a forest ranger. The chances were that he would have forgotten all about it the following day.

Sam was a hard worker and we now had the cabin completed to the stage where we could move in. We had been comfortable enough in the tent, but it was a bit cramped.

A Long Foot Patrol

It was now time for Sam to go into town, but by this time the river had dropped considerably so he would have no difficulty in fording. I had been longing to do a bit of exploring and travelling but I hesitated for two reasons: first because it was too hot for hiking and second because our meagre supply of fire fighting tools was back at Rock Lake. The fire hazard had reached the extreme stage and only a thunderstorm was needed to start a rash of forest fires.

I buried myself in work, hewing out a door and peeling the logs inside and out. The logs should have been peeled before they were used for building, but as I mentioned before, the cabin had already been started before we took over. Smoke began to drift in and it became so dense that visibility was restricted to a few yards. I had visions of being burned out, sometimes imagining I could hear flames crackling. I had almost decided to take my horses and belongings down to the river when the weather finally broke. It commenced with a violent thunderstorm, followed by hail, rain and snow for three days without letting up and I was certainly glad that I had moved into the cabin.

The storm ceased just as suddenly as it commenced. The sun came out again quite hot and what a mess it made in melting about eight inches of snow. There was water everywhere and then the snow slides started in the mountains. It was a wonderful sight to watch. First you would notice a patch of snow high up begin to move. As it gained momentum, a sound like thunder reached you. It would spread out, sweeping boulders and trees ahead of it until it finally came to rest in the valley, the sound reverberating from mountain to mountain.

I started out hiking as I did not want to take the horses along. I was not sure where I would find feed for them and besides I might want to do some mountain climbing.

Jack Glen in the mountains, c. 1920s.

JACK GLEN PHOTOS, ALBERTA FOREST HISTORY PHOTOGRAPHIC COLLECTION

When I reached the rock that the eagles had built their nest on, I looked it over carefully with my binoculars and figured that the nest could be reached by following a narrow ledge around the face of the cliff. This was too dangerous to attempt on my own just for the sake of a photograph. I had consulted the map issued by the Forest Service and the route I planned to take did not appear to be much over thirty miles. First I would travel to Little Grave Flats. From there I would leave the Mountain Trail, go north through what I hoped was a pass and climb to the headwaters of the Wildhay and Little Berland rivers. Somewhere in that locality I should find a sort of trail leading down the Wildhay to a spot opposite and about a mile from the Eagles Nest Cabin.

I travelled through the pass and the trail entered a long and lovely valley, with a creek known as Rock Creek meandering down the centre of it. Here were the haunts of meadowlark and ptarmigan. Moose and deer were also in evidence and I could see a bunch of bighorn sheep, no doubt watching me, from a rocky ledge high up. They have wonderful eyesight, but more about that later.

I was now close to the north boundary of Jasper National Park. Cairns on the mountaintops marked the boundary between it and Athabasca Forest Reserve. I reached the height of land between Rock Creek and the Sulphur River valley and ate my lunch. Later we built the Mile 58 Cabin close to that spot.[8]

Continuing on my way I arrived at the Little Grave. An Indian baby[9] had died on the trail and had been buried there, hence the name. It was fixed up in the usual Cree custom with a roof over it and a fence built around it. Years later, a lone traveller seeking shelter on a stormy night had found it and slept there. Nearby I found a trapper's cabin and decided to stay there for the night. An old rusty stove would give me warmth and boil water for my tea. At an altitude of close to 5,000 feet, the nights get chilly and I was packing only one blanket. I cut some spruce boughs and put them on the bunk. A little creek flowed past the cabin and I was successful in catching two nice Dolly Varden trout. I had no frying pan so I decided to boil them, but my supper was not altogether a success. The fish without salt tasted rather insipid and despite several rinsings of the pot in which I boiled them before using it to boil water for my tea, the tea tasted fishy.

I looked forward to a good sleep that night but I had no sooner dropped off than the mice started to serenade me. They ran all over me, squeaking, and what must have been a mother mouse started nibbling at my hair in search of material to build a nest. I armed myself with a stove wood stick, but just when I had almost dropped off to a sleep a timber wolf howled quite close to the cabin. I almost hit the roof and from then on sleep was impossible. I went breakfastless that morning, as the mice had destroyed what little food I had left.

I left the cabin at daybreak while it was still chilly, but by the time I reached the summit the sun was beating down mercilessly and I was perspiring freely. I sat down to rest and look around me.

[8] Locations were often identified by the distance by trail from the Entrance headquarters. Mile 58 Cabin (also called Summit Cabin) was 58 miles (93 kilometres) from Entrance.

[9] The grave is identified as that of "Baby Delorme, 1908" in the new Willmore Wilderness Park Map (June 2007), Alberta Tourism, Parks, Recreation and Culture.

Opposite page: Last of the Legend. *Woodland Caribou print by John Hiebert, 1989.*

JOHN HIEBERT, HINTON TRAINING CENTRE COLLECTION (ORIGINAL IS IN COLOUR)

Little Grave Flats – Fay Expedition 1914.

FAY EXPEDITION, JASPER YELLOWHEAD MUSEUM AND ARCHIVES JYMA 84.87.52

What a lovely panorama presented itself. I looked down on two lovely valleys. To the north was the valley of the Sulphur, which I had just left, and I could almost see to where it entered the Smoky River. Looking up-river I could see Hardscrabble Pass and beyond the pass, the Rockies showed up magnificently with Mount Bess topping all the neighbouring mountains. Mount Robson was not visible from that spot. To the north, the country appeared very rugged, the source of many creeks and rivers. I soon spotted the valley of the Wildhay down which I would travel to reach Eagles Nest.

At the summit I noticed several little rock rabbits or conies and wondered how such small delicate-looking animals could survive the rigors of winter at altitudes of 8,000 feet and more. Ptarmigan rose from almost under my feet and flew cackling away. I figure that they rely on their plumage as camouflage because they would be almost invisible from the air and that is where most of their danger lies. They change colour twice in a season. In summer their plumage is golden, flecked with black bars, and in fall and winter it turns white. All of this I had expected to see, but I was not prepared for what happened next.

I had now begun to feel tired and somewhat sleepy after my sleepless night and strenuous climb. My ears were drumming from the altitude, so I decided to sleep for a time. I walked over to a rock outcropping so as to be in the shade, when suddenly six animals jumped up and began snorting and milling around. From pictures I had seen, I recognized them as Osborne caribou,[10] much larger than the barren land caribou or a mule deer.

[10] Today's woodland caribou were once known as Osborne caribou.

Opposite page: Map 3. Jack Glen's First Foot Patrol, 1920. The likely route traversed from Little Grave Flats to the Indian Trail was interpreted from topographic maps.
FOOTHILLS RESEARCH INSTITUTE HISTORY SERIES

The bull caribou is a magnificent animal with his huge antlers and a neck shining like silver. They are given to curiosity and their antics scared me considerably. First they uttered a snort, almost a whistle, then they wheeled and dashed away for a distance of about a hundred yards. Then they came charging back and ploughed to a sudden stop about seventy-five feet from me. As this was my first encounter with caribou I admit that I was thoroughly scared. Every time they charged I figured that I would be trampled and perhaps gored to death.

I learned later that they performed these antics in order to get the scent of either man or beast and they relied on their speed to get away from danger. At last they seemed to be satisfied, for after coming in from a different direction they stopped as usual, snorted, wheeled, but did not return, much to my relief. I have since seen some excellent movies taken by big game hunters of the caribou antics and the final coup de grace.

The encounter with the caribou had banished all thought of sleep and now all I wanted was food and the solitude of Eagles Nest Cabin. I started out briskly and soon found a sort of trail that led down to the Wildhay River.

Darkness was coming on and I figured that by crossing the river and angling in the direction of the cabin I would save myself some mileage. This is where I proved the adage that the longest way round is the shortest way home, for no sooner had I crossed than I found myself in a long open muskeg. It was now pitch dark and I kept falling into holes.

I worked my way out to the fringe of the timber, but although I did reach dry ground I stumbled over roots and was in danger of having my eyes knocked out, so I had to go back to travelling in the swamp again. I was soaked to the skin and plastered with mud from head to foot.

Around midnight I finally arrived at the cabin and figured afterwards that the round trip was much closer to forty than thirty miles. I found that Sam was back from his trip to the outside but he was asleep so I did not wake him. I ate a huge meal, stripped off my wet and muddy clothes and got into bed. For the next fifteen or sixteen hours, Rip Van Winkle had nothing on me.

Some days later I decided that I would like to see a little more of my district, this time with my horses along. Although the cabin was yet unfinished, Sam and I agreed it could wait until the weather was less agreeable. As it turned out, it would be years before the job was completed. We started off in different directions. Sam wanted to see the headwaters of the Big Berland and Muskeg rivers. I followed my previous route and camped on the summit where I had eaten lunch. I pitched my tent and caught fish enough for supper and breakfast in Rock Creek. The meadowlarks were still singing their song of praise and I was awakened in the morning by the crowing of a ptarmigan.

I discovered another Indian grave, this one in a sad state of neglect. The skull and bones were lying on the surface of the ground and a crude fence of logs surrounded the skeleton. Later I asked the Indian who trapped that district about

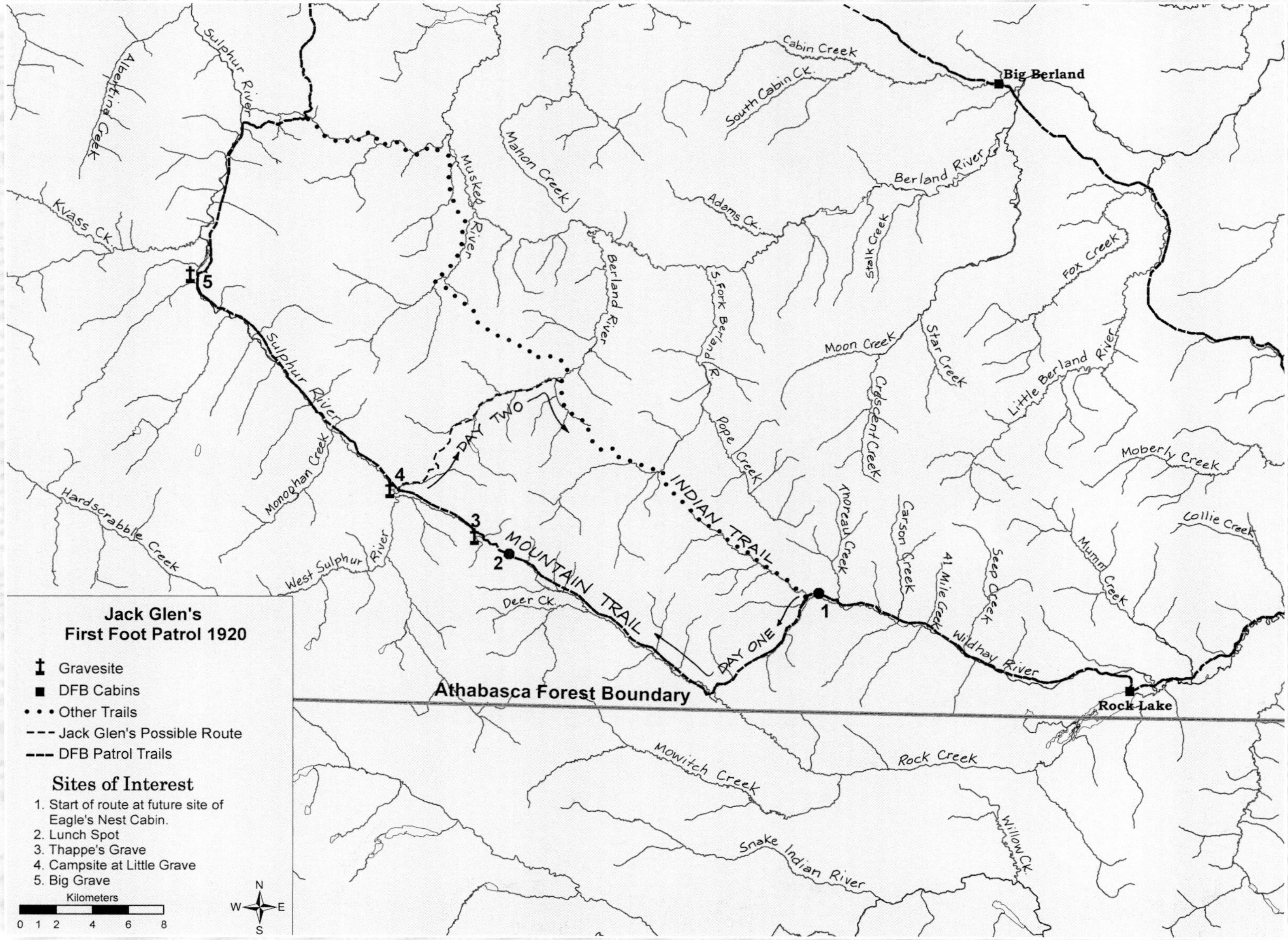
Jack Glen's
First Foot Patrol 1920
Gravesite
DFB Cabins
Other Trails
Jack Glen's Possible Route
DFB Patrol Trails
Sites of Interest
1. Start of route at future site of Eagle's Nest Cabin.
2. Lunch Spot
3. Thappe's Grave
4. Campsite at Little Grave
5. Big Grave
Kilometers
0 1 2 4 6 8
N
W
E
S
Big Berland
Rock Lake
Athabasca Forest Boundary
MOUNTAIN TRAIL
INDIAN TRAIL
DAY ONE
DAY TWO
Cabin Creek
South Cabin Ck.
Berland River
Adams Ck.
Stalk Creek
Fox Creek
Mahon Creek
Musket River
Berland River
S. Fork Berland R.
Moon Creek
Star Creek
Little Berland River
Crescent Creek
Pope Creek
Moberly Creek
Collie Creek
Thoreau Creek
Carson Creek
41 Mile Creek
Seep Creek
Mumm Creek
Wildhay River
Sulphur River
Albertina Creek
Kvass Ck.
Sulphur River
Monoghan Creek
Hardscrabble Creek
West Sulphur River
Deer Ck.
Mowitch Creek
Rock Creek
Snake Indian River
Willow Ck.
1
2
3
4
5

Sam Nilson at Big Grave Flats, 1947-48, Art Allen's outfit.

MARK TRUXLER COLLECTION, COURTESY OF JACKIE HANINGTON

Grande Cache

The families in Grande Cache did not receive clear rights to their lands until the 1960s when James Shand Harvey wrote a statement asserting that he was present when government commissioner J.W. McLaggan told the Jasper Métis families that if they vacated their lands in the newly formed Jasper National Park, in addition to the cash settlement, they could move and settle on vacant Crown land anywhere they wanted outside the Jasper Park boundaries. Shand Harvey's affidavits confirming those events helped legitimize their ownership of the lands in the Hinton and Grande Cache areas where they settled.

Sources: MacGregor, J.G. (1962). Packsaddles to *Tête Jaune Cache*, Hurtig Publishers. pp. 227–229. Interview with James Shand Harvey by Mr. and Mrs. Mark Truxler, 8 March 1967, Jasper Yellowhead Museum and Archives.

the details. He told me the story rather unwillingly, because they do not like to discuss these things. Nor do they care to have their graves photographed, but I took a snap of it anyway. A strange tribe had been travelling through there and one of the party had died. It was late in the fall and the ground was frozen so hard that burial was out of the question. They built a fence around the body and piled stones on it and departed, meaning to come back again in the spring and plant him properly. They never got around to doing it.

Years later I was showing the grave to a hunting party and one of the hunters who hailed from Indiana was greatly interested in the story. He drew my attention to the state of the skeleton's teeth. "You'll note," he said in his southern drawl, "how the teeth are worn down in places. That, no doubt, comes from chewing on hard flinty squaw corn." I had a hard time keeping a straight face at this post-mortem, as I was pretty sure that the old boy's closest contact with "cawn" was the picture on the label of an empty corn can.

I followed down the valley of the Sulphur River, which is fed by many sulphur springs. I camped that night on Big Grave Flats. The grave there had been fairly well looked after. Tourists and hunters had carved their names on the fence rails and the mountain gophers had their burrows all around and under the grave.

The next morning I noticed a coyote trotting across the flats. He was giving my camp a fairly wide berth and all of a sudden he spied a gopher who had strayed a little too far from his burrow. With one mighty leap he had the gopher in his jaws, but did not stop to eat him as he was still suspicious of me.

These gophers make choice morsels for the grizzlies, and I have watched them digging them out. When Bruin goes into action he loses no time. A sniff at the mouth of the burrow indicates whether or not the gophers are home. If he has the green light the sod begins to fly, also the rocks, and in no time he will have a hole two feet deep, pausing in his efforts only long enough to sniff and listen to find out if he is on the right track. The digging is then resumed and the kill made. I have seen rocks that must have weighed at least a thousand pounds tossed aside like so many pebbles.

I was treated to another rare sight that morning. A beautiful bull elk appeared on the meadow and when he was nearly opposite my camp he threw up his head and bugled. That day it seemed as if nature was putting on a show especially for my benefit.

After lunch I walked down the trail. This is the turning point on the Mountain Trail. The trail follows along the river until you reach the Indian settlement of Grande Cache. You are then on the Lower Trail that leads back to Entrance. The Indians there, and in fact nearly all those living in the Athabasca Forest Reserve, originally roamed what is now Jasper Park. When the Dominion Parks Board decided to create the Park they paid the Indians a sum of money to vacate. This they did, but they did not go far, just squatting on the choicest spots they could find in the Forest Reserve and there they remained despite all efforts of the provincial government to move them.

The Grande Cache tribe brought in cattle and also packed in a hay mower and rake. They have bred some fine saddle and pack horses. The men act as guides in the summer and fall and the women tan the hides of the game animals and turn out some lovely bead work on buckskin coats, gun scabbards, gloves and moccasins. In later years I had to depend on these people to fight our forest fires.

During my walk that day I disturbed a huge porcupine gnawing at a shed moose horn. He waddled ahead of me for quite a distance, watching me out of the tail of his eye. When I got close to him up came his quills and he contrived to keep his rear towards me at all times as the head and belly are his two vulnerable points. It is said that the wolverine and the fisher are the only two animals that can kill and eat a porcupine and live. I shot a bighorn ram once that obviously had had a bout with one. He had quills sticking in his forehead right into his skull. I imagine that he killed the porcupine, as I have watched two rams fighting, and the impact when they butted horns was terrific. Dogs sometimes tackle porcupines and unless all quills are pulled out it may mean death for them. I finally bypassed the creature and walked on.

Next I watched an otter amusing himself. He would climb up a steep clay bank of the river, then slide down into the water and cavort around in a deep pool for a time. He did this several times while I was watching. The slide that he created was perfectly smooth and each time he went down, the water from his fur lubricated the slide and made it very slick indeed.

Later, as I focussed my binoculars on the mountain, soon across the river I located a salt lick with a number of sheep sampling it. Suddenly they appeared to take fright. Searching for the cause I finally located what I took to be a wolverine, but his chances of killing a sheep would be most remote. Sheep are the most wonderful climbers under the sun. Some people say that the mountain goat is the best, but for my money I will say that the goat is not in the same class as the sheep. He is much slower and more awkward. He frequents more rugged terrain and never have I known the two to mingle.

The mountain goat is a browser and will generally be found well above the timberline in summer, but the snow in the fall will drive him to the timber where he winters and sheds his old coat of hair in the spring. Mountain sheep prefer the grassy southern slopes of the mountains and are also driven down to the timber in the fall where they eke out a precarious existence, emerging in the spring more dead than alive. They gain weight rapidly as the succulent grass on the mountain slopes grows more quickly and seems to be very nourishing. The sheep's meat I think cannot be beaten as far as game animals go.

You can easily outsmart a goat, but not a sheep. When a goat no longer sees you, he relaxes and can be stalked unless he is in a spot inaccessible to you. The

Bighorn Sheep graze in meadows above Jasper Lake.
CHARLES KAY

sheep are wary and always have a sentinel, generally an old ram, stationed on a point of rock. If you try to stalk them from below he will give the alarm and the whole bunch will be over the summit in no time. They never seem to expect danger from above, where you can get to with a long detour and some hard climbing. When you start to shoot they will still climb – unless one is too badly wounded to climb – for they seem to think that safety lies in height.

I spent a few more days at this camp and climbed some of the lower mountains, as I wanted to familiarize myself with the existing water supplies in case of forest fires.

The Grave Flats Grizzly

I had heard various guides talk of a monster grizzly that frequented the Grave Flats country. He was an ugly brute, they said, and I could believe it after I saw what he had done to a packhorse. The poor beast had been clawed to the bone and was a mass of festering sores. I decided to hunt and if possible kill this brute. He would make a wonderful trophy and I would be ridding the district of a pest. I started out early one morning and hunted him all day. When I started back to camp the light was failing and it began to snow. I had lost my enthusiasm, besides which I did not relish the idea of spending a night in the open with a killer grizzly on the prowl. Suddenly I ran across his fresh tracks in the snow and for a moment I imagined I was gazing at the tracks of some prehistoric monster. I have never since seen tracks to compare with those. I reached camp all right, but I was pretty

jittery and I honestly believe that if a ptarmigan had risen out of the buck brush under my feet I would have had a heart attack. I have often thought since how fortunate I was not to come face to face with him on that occasion as I was totally inexperienced in grizzly hunting and my rifle was much too light. I had heard stories of the cunning and ferocity of grizzlies before, but thought them greatly exaggerated. In later years, numerous encounters convinced me that he was not to be taken lightly.

Two men whom I got to know very well went gunning for the old renegade shortly after this, Jack Brewster and Felix Plante, and they were considered among the best of hunters and guides at that time. They finally located the bear and caught a fleeting glance of him. Brewster fired but lost sight of him among the buck brush. Sneaking over carefully to the spot where he last saw him, he found blood in a considerable quantity and started to track him. In the meantime Plante had been watching from a vantage point and suddenly saw that the hunter had become the hunted. A well-aimed shot from his rifle ended the hunt. These men knew their grizzlies. Had they both taken after the bear, one or possibly both of them might easily have been killed when he sneaked up behind them.

It also goes to show, the cunning of the grizzly. If he had charged Brewster when he was hit, he no doubt instinctively reasoned that he would be killed. When hit in the open a bear will either play possum or charge. I have known them do both, but the Brewster bear had good cover and took advantage of it.

Jack Brewster with Trophy Sheep head, 1918.

JASPER YELLOWHEAD MUSEUM AND ARCHIVES
JYMA PA 34-17

I am a great horse fancier and when I saw Mr. Riviere's bunch I could hardly take my eyes off them. They were short-coupled and sturdy, real horses for a mountainous country. He saw me admiring them and explained that they were his utility bunch, which he had broken personally.

Forester Charles H. Morse with his pack outfit on Glacier Creek near Mount Bess, Athabasca Forest, c. 1914.

DOMINION FORESTRY BRANCH, ALBERTA FOREST HISTORY PHOTOGRAPHIC COLLECTION

CHAPTER TWO

TRAVELS WITH FRENCHY RIVIERE

It was almost time to make another trip to Entrance as my food supply was getting low and my diary had to go in. Back at Eagles Nest I laid out some trail work I wanted Sam to attend to while I was gone. He told me confidentially that Munson was an assumed name, which he had adopted when eluding the sheriff in Montana, no doubt. I learned his real name but have never divulged it. When I said good-bye to him there at Eagles Nest, little did I think that I would never see him again.

On my way out, at Moberly Creek, I called Mr. Badgley. He first enquired about the state of my health and how the rainbow fishing was, then told me he had a lovely trip lined up for me if I was interested. It would be entirely voluntary on my part and I could refuse it if I wished to. He had a letter from the Chief Provincial game guardian, a Mr. Riviere, who wished to make an inspection of our Forest Reserve and the Jasper National Park with a view to taking a census of the game with the cooperation of the forest rangers and park wardens, and straightening out some grievances over traplines. He had tried to hire a helper in Edmonton but could not find anyone suitable, so in desperation he had written Mr. Badgley to see if he could spare any of his rangers. Naturally, I jumped at the offer.

From Entrance, I contacted Mr. Riviere, who asked if I could supply my own saddle horse and meet him in three days at Brûlé, a busy mining town that had proper chutes for unloading his horses and equipment from the train. When I had unloaded at Entrance earlier that year, we had to build up a sort of ramp out of railway ties, planks and what have you to get my gear off the train.

Opposite page: Henri "Frenchy" Riviere outside Prince of Wales Hotel, Waterton Lakes National Park, Alberta, c. 1930. Riviere was a flamboyant character and his colourful dress style including the Métis sash around his waist exemplifies this.

GLENBOW MUSEUM AND ARCHIVES PA 3433-1

The mining Town of Brûlé in the early 1920s. The Blue Diamond Mine was the main employer in the town, established to provide coal to the Canadian Northern Railroad.

DENNIS RADCLIFFE PHOTOS, ALBERTA FOREST HISTORY PHOTOGRAPHIC COLLECTION

When the freight train stopped at Brûlé, Mr. Riviere alighted from the caboose. He was a very tall man, slightly stooped. His attire was a combination of sea pirate and western gold miner. He never wore a hat and I could not hazard a guess as to his age. His greying hair was hanging down his back, and around his neck he wore a red bandanna. Buckskin pants and shirt completed the outfit, with a gaudy sash in lieu of suspenders, together with a cartridge belt and a huge knife, which he referred to as a pig sticker, stuck down between his pants and sash. Incidentally the Mountie at Brûlé frowned upon the pig sticker and told him to get rid of it or keep it out of sight. That knife had many uses. He used it for buttering his bread, cutting his meat, removing slivers, trimming his horses' hoofs and tails and cutting his own finger and toe nails. A very versatile piece of equipment.

I approached him and introduced myself, joking that "I take it that Riviere will ride again." He made a sweeping bow. "At your service, sir," he said.

We soon became fast friends and our trip was most enjoyable from start to finish. I am a great horse fancier and when I saw Mr. Riviere's bunch I could hardly take my eyes off them. They were short-coupled and sturdy, real horses for a mountainous country. He saw me admiring them and explained that they were his utility bunch, which he had broken personally. He guaranteed that there was not a bucker in the bunch. This did not mean that they were deadheads, however, as they could travel at better than three miles an hour, considered good going for a loaded pack horse. He owned a horse ranch in Pincher Creek and raised horses for the Mounted Police, keeping his eyes open also for a promising polo pony. A real good polo pony, he noted, often gets to know the game better than his rider and therefore they fetch a high price. The bunch that he had with him, he explained, did not come up to Police standards, as they preferred a narrow-chested, "rangier" type.

On the Trail into Jasper National Park and the Athabasca Forest Reserve

We packed up and got started on our way. Brûlé at that time was within the boundary of Jasper Park so the Mountie sealed our firearms as the game therein was protected. We made just a short trip the first day, as the horses were tired after their long trip in the stock car. It was a lovely campsite there by Moose Creek[11] and Mr. Riviere elected me cook, which suited me fine, as horse wrangling can be miserable business at times. His cooking outfit was scanty to say the least and was contained in a small canvas sack, which on several occasions resembled a porcupine when the knives and forks broke loose from their moorings. A large Dutch oven was included, together with a ten-pound lard pail, which served for brewing tea, coffee or at times chocolate. But what he lacked in cooking utensils he certainly made up for in food.

[11] Moose Creek was renamed Moosehorn Creek about this time.

The first thing I did, after unpacking, was to pick up the axe to chop fire-

Frenchy Riviere

Henri "Frenchy" Arnous de la Riviere (1867–1956) was born in Normandy, France. He received some education in Switzerland but ran away to sea at an early age. He later moved with his parents to the southern United States. He travelled throughout the western USA and arrived in Alberta in 1883 as a cowboy. He was a guide during the Klondike Gold Rush and afterwards settled on a homestead at the headwaters of Pincher Creek. He worked as a big game hunter and, with John Herron as a partner, raised and broke horses and raised husky dogs. His knowledge of dog teams was used in a Hollywood film. Riviere also wrote many wildlife stories, some of which were published in magazines such as *The Cattleman.*

Frenchy Riviere training his dog.
GLENBOW MUSEUM AND ARCHIVES NA - 184-42

From 1911 to 1928 he was employed as an Alberta game guardian in northern Alberta and in that capacity travelled widely from his home base in the Pincher Creek area. We know from archival correspondence that he travelled to Grande Cache in 1917 to comment on the Métis settlers (he supported their being there) and in 1918 he was along the Brazeau to comment on the location of the south boundary for Jasper Park. He and his wife, Nellie Gladstone (d. 1940), had 11 children, Henry, George, Mary (Shriver), Nellie (Murphy), Robert, John "Charlie," James, Alice (Primeau), Inez (Rea), Frances (Cox, then later McWhirter), and Emily.

His relationship with the Dominion Forestry Branch was not always cordial. Department supervisor Ward Badgley, in a 1917 report, was alarmed that Riviere and his associate Thomas Groat appeared to have encouraged the Grande Cache natives to stay on their lands despite efforts by the Department to remove them. In light of this and his concerns that further native settlement could occur north of Grande Cache, Badgley recommended immediate expansion of the "territory" – presumably the Athabasca Forest – to the north.

Badgley viewed Riviere as somewhat high-handed and a threat to the Department's interests in other areas, for example, expanding the DFB patrol trail system. Badgley's 1917 comment about Riviere suggests a mutual antipathy: "Revere [sic] says there is not a trail or cabin in the Reserve fit to stop in or travel over, all pure waste of Government money. All the old Indian trails are here yet and strange to say Revere travels the Gov[ernment] roads and makes his boasts that he is furnished keys of all Forestry cabins in the Rocky Mountain Reserves and has authority from Ottawa to use them and report general conditions in supervision."

Sources: Glenbow Museum Archives, Calgary, Alberta; Southern Alberta Pioneers www.pioneersalberta.org/; and *Riviere, Henri Arnous*, 1868–1956, Sikina Publ., Calgary, 1997.

Badgley reference and quotation courtesy I.S. MacLaren - Library and Archives Canada, RG 39, vol. 294, file 42489, W.W. Badgley to R.H. Campbell (Dominion Superintendent of Forestry), 18 Oct. 1917, 4 pages typescript.

Henri "Frenchy" Riviere in camp c. 1940s.

GLENBOW MUSEUM AND ARCHIVES NA 3051-3

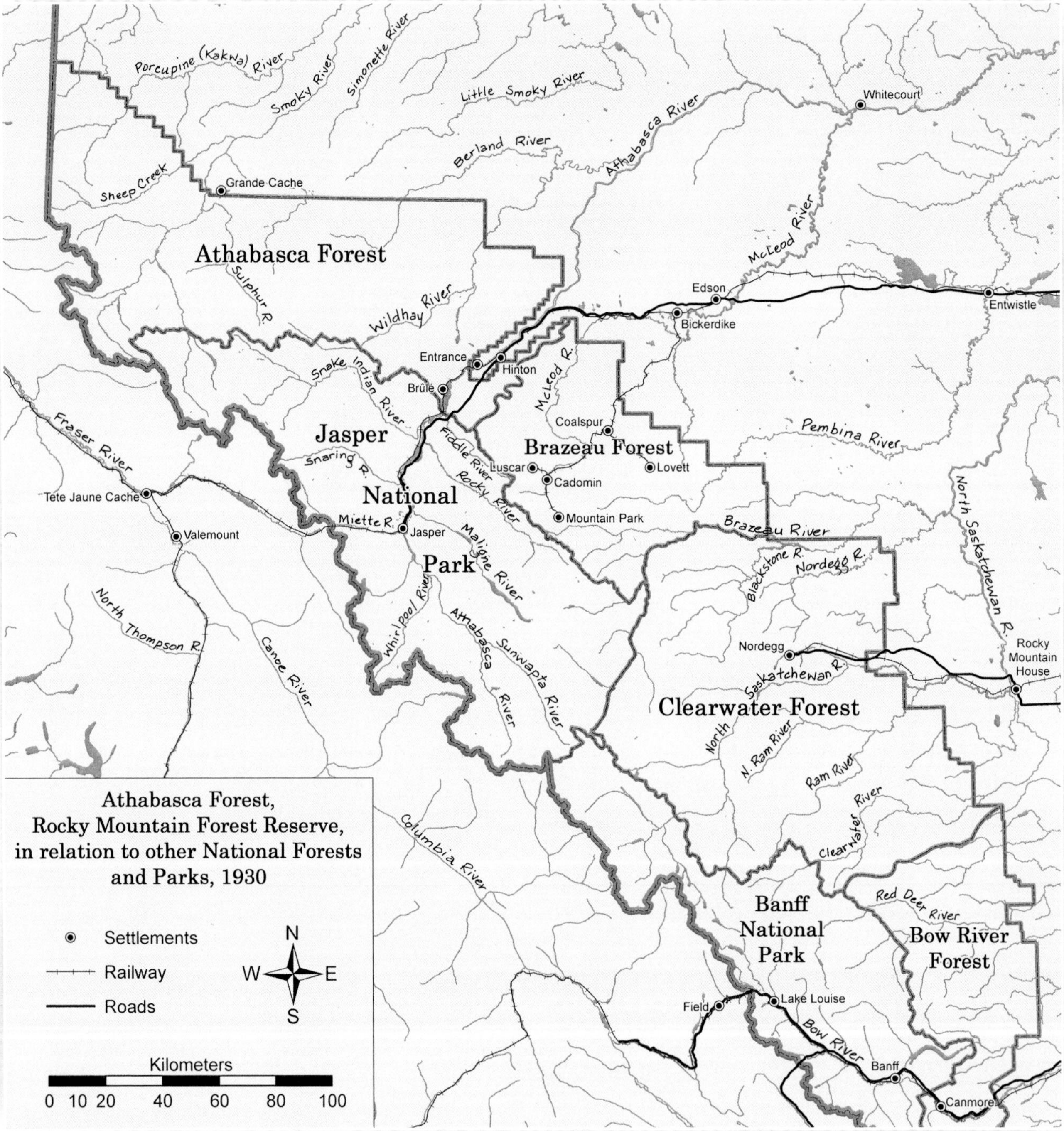

Map 4. Athabasca Forest in relation to Jasper National Park and the other forests of the Rocky Mountains Forest Reserve, 1930.

wood. What an axe that was! The edge, without exaggeration, was at least one-sixteenth of an inch thick. I brandished it in front of Mr. Riviere and asked him if by any chance he had ever considered introducing it to a grindstone.

He grinned, and offered his reason for being dilatory in the care and upkeep of an axe: "Friend," he said (he always addressed me as Friend), "in the winter I do considerable dog mushing and when you hit camp, cold and hungry, I don't know of a better way to warm up and start your blood circulating than chopping wood with a dull axe."

After supper I asked him if he happened to have a file in his outfit. He had and I went to work on the axe. To my way of thinking I would much sooner chop my wood with a sharp axe and do my warming up afterwards by a rousing campfire.

Outside of neglecting his axe, Mr. Riviere was a wonderful woodsman and I learned a lot from him during the trip. He told me the story of his life, which helped me understand his nickname – "Frenchy" Riviere. His parents were old country French who had settled somewhere south of the Canadian border. He was orphaned when quite young and raised by friendly Indians who were continually on the move. They taught him their customs and he learned how to use the bow and arrow, and later the muzzle-loading flintlocks. He wore a breech clout just like his young native playmates and spoke the Cree language just as well as a native.

This knowledge was very useful. At that time, the Dominion Forest Service in Alberta was in its infancy and had printed a generous number of posters setting forth Forestry regulations, warning most important of all to campers the necessity of putting out their camp fires before leaving. These posters were put up in prominent spots such as river fords, and picnic and camp sites. Now a great number of Indians travelled the trails and, if they noticed the white man's posters, they meant nothing to them, as most could not read English. Then someone suggested the use of posters in Cree. Many white men could speak and understand the language, but try and find one who could write it. Mr. Riviere and a Mr. Harvey[12] were asked to do it. A Cree poster looks like something out of this world with its symbols, loops and whorls, triangles and characters, but the important thing was that it got the message across to the Indians.

We were not familiar with this part of the country, and maps at this time were not too accurate, so whenever we came to a spot that would make a good campsite, with wood, water and grass in abundance, we camped. I preferred a new campsite anyway, as a much-used camp was generally infested with mice, pack rats, rabbits and porcupine.

We used a teepee, which is more comfortable than a tent, providing you can eliminate the smoke. To do this you must understand how to manipulate the wing at the top that acts much the same as a damper on a stove. Native women seem to be the only ones who can adjust it properly. A pole from the ground to a

[12] James Shand Harvey was a long-time and well-known resident of Entrance. He was born in 1880 on the Island of Mauritius, attended Eton, and was commissioned in the British army. He came to Canada in 1905 to see the country and stayed, working on the 14th Base Line Survey, and became a guide and outfitter, trapper and forest ranger. His home was at Entrance and he was the subject of historian J.G. MacGregor's book *Pack Saddles to Tête Jaune Cache* (Hurtig Publishers, 1973). Annual reports show him listed as a forest ranger in 1916 and 1917 he may also have worked through to 1924. Shand Creek was named after him. (Hazel Hart, *History of Hinton*, 1980; Aphrodite Karamitsanis, *Place Names of Alberta*, 1991; James G. MacGregor, *Packsaddles to Tête Jaune Cache*, 1962)

Prevention of fires in the Forest Reserves was a major objective and the DFB obviously took the task seriously. This fire prevention notice is written in Cree syllabic.

DOMINION FORESTRY BRANCH, DEPARTMENT OF THE INTERIOR, ANNUAL REPORT, 1910

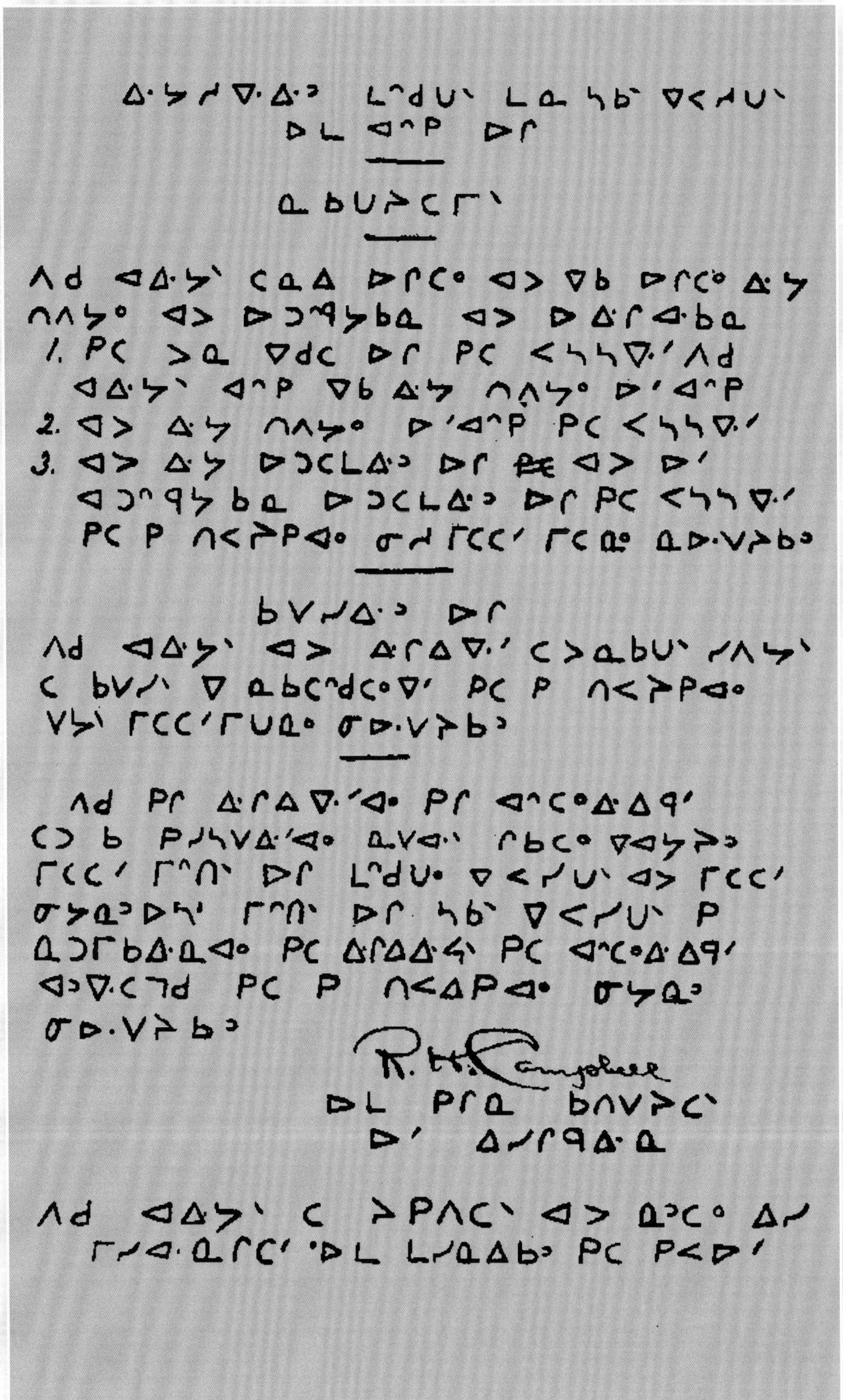
ᐃᐧᓭᓱ ᐁᐧᐃᐧᐣ ᒪᐢᑯᑌᐠ ᒪᓇ ᓴᑳᐠ ᐁᐸᓯᑌᐠ
ᐅᒪ ᐊᐢᑭ ᐅᒋ

ᓇ ᑲᑌᔨᑕᒥᐠ

ᐱᑯ ᐊᐃᐧᔭᐠ ᑕᓇᐃ ᐅᒋᑕᐤ ᐊᐳ ᐁᑲ ᐅᒋᑕᐤ ᐃᐧᔭ
ᑎᐱᔭᐤ ᐊᐳ ᐅᑐᐦᑫᔭᑲᓇ ᐊᐳ ᐅᐃᐧᒋᐊᐧᑲᓇ
1. ᑭᑕ ᐳᓇ ᐁᑯᑕ ᐅᒋ ᑭᑕ ᐸᓴᓴᐁᐧᐟᐱᑯ
ᐊᐃᐧᔭᐠ ᐊᐢᑭ ᐁᑲ ᐃᐧᔭ ᑎᐱᔭᐤ ᐅᐟᐊᐢᑭ
2. ᐊᐳ ᐃᐧᔭ ᑎᐱᔭᐤ ᐅᐟᐊᐢᑭ ᑭᑕ ᐸᓴᓴᐁᐧᐟ
3. ᐊᐳ ᐃᐧᔭ ᐅᑐᑕᒪᐃᐧᐣ ᐅᒋ ᐊᐳ ᐅᐟ
ᐊᑐᐢᑫᔭᑲᓇ ᐅᑐᑕᒪᐃᐧᐣ ᐅᒋ ᑭᑕ ᐸᓴᓴᐁᐧᐟ
ᑭᑕ ᑭ ᑎᐸᔨᑭᐊᐤ ᓂᓱ ᒥᑕᑕᐟ ᒥᑕᓇᐤ ᓇᐅᐧᐯᔨᑲᐣ

ᑲᐯᓯᐃᐧᐣ ᐅᒋ
ᐱᑯ ᐊᐃᐧᔭᐠ ᐊᐳ ᐃᐧᒋᐃᐁᐧᐟ ᑕᐳᓇᑲᑌᐠ ᓯᐱᔭᐠ
ᑕ ᑲᐯᓯᐠ ᐁ ᓇᑲᑕᐦᑯᑕᐤᐁᐧᐟ ᑭᑕ ᑭ ᑎᐸᔨᑭᐊᐤ
ᐯᔭᐠ ᒥᑕᑕᐟᒥᑌᓇᐤ ᓂᐅᐧᐯᔨᑲᐣ

ᐱᑯ ᑭᒋ ᐃᐧᒋᐃᐁᐧᐟᐊᐤ ᑭᒋ ᐊᐢᑕᐤᐃᐧᐃᑫᐟ
ᑕᐳ ᑲ ᑭᓱᓴᐯᐃᐧᐟᐊᐤ ᓇᐯᐊᐧᐠ ᒋᑲᑕᐤ ᐁᐊᐧᔭᔨᐣ
ᒥᑕᑕᐟ ᒥᐢᑎᐠ ᐅᒋ ᒪᐢᑯᑌᐤ ᐁᐸᓯᑌᐠ ᐊᐳ ᒥᑕᑕᐟ
ᓂᔭᓇᐣᐅᓴᐟ ᒥᐢᑎᐠ ᐅᒋ ᓴᑳᐠ ᐁᐸᓯᑌᐠ ᑭ
ᓇᑐᒥᑲᐃᐧᓇᐊᐤ ᑭᑕ ᐃᐧᒋᐃᐃᐧᔭᐠ ᑭᑕ ᐊᐢᑕᐤᐃᐧᐃᑫᐟ
ᐊᐣᐁᐧᑕᒣᑯ ᑭᑕ ᑭ ᑎᐸᐃᑭᐊᐤ ᓂᔭᓇᐣ
ᓂᐅᐧᐯᔨᑲᐣ

R. H. Campbell
ᐅᒪ ᑭᒋᓇ ᑲᑎᐯᔨᒋᐠ
ᐅᐟ ᐃᓯᑫᐃᐧᓇ

ᐱᑯ ᐊᐃᐧᔭᐠ ᑕ ᔨᑭᐱᑕᐠ ᐊᐳ ᓇᐣᑕᐤ ᐃᓯ
ᒥᓯᐊᐧᓇᒋᑕᐟ ᐅᒪ ᒪᓯᓇᐃᑲᐣ ᑭᑕ ᑭᐸᐅᐟ

Translation:

PRAIRIE AND FOREST FIRE LAWS FOR THIS COUNTRY

Take Notice

Any one who purposely or not purposely or his servant or his companion:

(1) Makes a fire and allows it to run on anybody else's land not his own, or,

(2) Allows a fire to run from his own land, or,

(3) By his doing or by his servant's doing, allows any fire to run, will have to pay $200.

About Camping

Any one, or any one that is with him, who lights a fire in the open for camping, and leaves without putting it out, will have to pay $100.

Those who must go and help put out a fire.

Those who are grown up and have not reached 60 years of age, being within 10 miles of a prairie fire or 15 miles of a bush fire, and being called upon to help put it out, refuse to help, will have to pay $5.

(Sgd.) R. H. CAMPBELL
The one who is the head of this work.

Department of the Interior,
Ottawa, 1908.

Any one who tears down or destroys this notice will be put in prison.

pocket in the wing is the controlling agency. I never became really expert, but managed after a number of attempts to get by with the minimum of smoke.

We slept with our feet to the fire and were really comfortable, especially on stormy days when we didn't travel. I always made sure to have lots of dry wood and kindling by the side of my bed, so that I could reach out and start the fire in the morning without getting up.

We travelled at a leisurely pace through Jasper National Park and the scenery was wonderful, especially when we were in the Snake Indian River Valley, also known as the Stony River. Princess Creek, which empties into the Stony, was the site for a warden's cabin, then in the construction stage. We decided to camp there as we spied a huge salt lick on the mountainside across the river.

Upper valley, Snake Indian River, looking north, 2003.

BOB UDELL

The next morning we were treated to a wonderful sight. I believe every breed of animal that frequented Jasper Park, with the exception of bear, was represented there. Elk, goat, caribou, deer, moose and sheep – which were in the majority – were all there and wallowing in that black gooey mess. Game trails like the spokes of a wheel converged on the lick from all directions. Sheep especially are crazy for salt, so much that the warden at Brûlé used to pack stock salt up the mountain to keep them from coming down to the railroad in search of leftover food thrown from the dining car.

Snake Indian River upstream from the site of the old Devona Warden Station, 2003.

BOB UDELL

Years later I was taking my wife and son up a steep mountain road to the Miette Hot Springs. Rounding a bend in the road we ran onto a flock of ewes and lambs busily engaged in tearing a large cardboard carton to pieces. It no doubt had contained something salty. I hated to disturb them but we wanted to be on our way. I tooted the horn and they ran ahead of us showing no inclination to leave the road. As we edged closer they began to get agitated and started to climb up a steep cutbank. The rocks and gravel came rolling down on us and I expected to have a window broken any minute. Luckily we escaped without a scratch.

When we arrived at the springs, we found a regular tent town with mountain sheep rooting through the garbage everywhere. I believe miners from the Pocahontas coal mining camp originally discovered these springs, dug them out and walled up the banks with logs.[13] At that time there was only a pack trail leading to them, which the miners also built, so little camping was done. When the Dominion government built the present beautiful bath, sweathouses and highway, they cleared a large campsite. It was then that the sheep found their way down from the adjoining mountains. They soon made it their home and became lazy with no thought of rustling for themselves. The result was inevitable. When the hot springs was closed for the winter and their meal ticket was gone, they just hung around and many died of starvation. The violation of the laws of nature disrupted their normal way of living so much that they shed their hair at the wrong time of year and their mating season also went haywire.

[13] Ewan Moberly is generally credited with discovering the springs in the late 1800s while hunting sheep, but miners built the first log pool. Source: P.J. Murphy et al., *A Hard Road to Travel*, Foothills Model Forest, Hinton, 2007.

On the trip with Mr. Riviere, good grazing for the horses was always the first consideration in picking a campsite, and if we found a good one we often laid over for a day or two. This suited me fine as it gave me a chance to get ahead with my baking. When we left Brûlé we had a dozen or so loaves of bread, but two hungry men on the trail can eat a lot of bread. I used the Dutch oven for baking bannock. I would first get a good bed of glowing coals in the campfire, and then I would mix my batter, grease the Dutch oven, and dump the batter in, cover it with the heavy lid, and bury it in the coals. It could then be forgotten until the fire burned itself out. Once in awhile I baked a cake by the same method. I would have a separate fire for cooking the rest of the meal.

Sometimes I took my binoculars and hiked out with Mr. Riviere, who would invariably lead us to where there was game. I wondered for a time if he could scent game, so I finally asked him how he always knew where the game was.

"Elementary, my dear Watson," he said with a grin. "Whenever you see eagles hovering about a certain spot, that's where you'll find game. If there's an osprey above water, there'll be fish for the taking."

On one occasion we watched a nanny goat defending her two kids. They were perched on a ledge with the two kids crowding as close as possible to their mother. Two eagles were attacking, but only one at a time had room to manoeuvre as the goat family were entrenched in a sort of alcove and, as they had their backs to the cliff, they were protected on both flanks and the mother had only the frontal attack to ward off. Every time an eagle dived she would meet it with a thrust of her horns and you could see feathers fly. The eagles were persistent but they finally acknowledged defeat.

Mr. Riviere and I continued our census and camped at such places as Little

Miette Hot Springs. The original pool was built with logs. c. 1930.

MOBERLY, HAZEL HART COLLECTION, HINTON LIBRARY AND ARCHIVES

Winnifred Lakes, near the headwaters of Kvass Creek, Willmore Wilderness Park, c. 2000.

SUSAN FEDDEMA-LEONARD AND THE WILLMORE WILDERNESS FOUNDATION

Heaven Summit, Winnifred Camp, and others, which if named I did not know. In the evening and on the days we did not travel, I fished and those lovely brook trout made a pleasant change in our menu.

We continued down the north fork of the Sulphur River and entered the Forest Reserve. The seals on our rifles could now be broken and, as it was September, the hunting season for sheep and goat was open. A special licence was needed to hunt any other type of game at that time.

He lost no time in killing a goat and, seemingly without effort, packed the entire animal, hide, horns and all, besides his rifle into a camp, a load of at least 150 pounds. He was truly the Canadian version of Daniel Boone, Davy Crockett and Buffalo Bill rolled into one. The meat was tender and juicy, for, as Riviere remarked jokingly, he had hand picked this young billy goat from a bunch of ten.

In the long evenings we lay on our beds by the warm fire, barbecuing spare ribs of goat and spinning yarns. I would make a pot of chocolate about eleven and sometimes roast potatoes in the hot embers. Then a bannock broken in two completed the feast. I would then roll into bed and fall asleep with Frenchy still talking.

Lessons in Horse Psychology

When I woke in the morning he would be gone. I don't know when that man slept. Before long I would usually hear the horse bells getting closer to camp, sometimes still in the dark. I then started preparing breakfast, as he liked a cup of coffee first thing. I once asked him why he went out after the horses so early, and he explained it this way:

"You know, Friend," he said, "the cayuse is quite a schemer. When you turn him out after a day's work, his first thought after he's had a roll is to fill his belly. This will generally take him till dark. Then he rests and in the meantime he's formulating a scheme whereby he can get out of work. Or maybe he's remembering a lovely patch of bunch grass or peavine or goose grass he and his companions found several camps back. His mates are probably thinking along the same lines, so he figures just as soon as it's light enough to travel, he'll be on his way. Right there is where I throw a monkey wrench into the works by interrupting his program and slipping a halter or bridle onto him. You see, he's much too wise to pass the camp because he knows he'd be caught."

I found out in time to come that besides his other accomplishments, Frenchy Riviere could evidently read a cayuse's mind. Many a time I have had murder in my heart when I went out to catch my horses. Some are really foxy and will let you get right up to them, but the minute you try to slip the halter or bridle on they start walking away and take a bit of grass now and then, trying to aggravate you further. About this time you are not in too happy a frame of mind, as you have walked already two or three miles and are soaked to the hide by the wet vegetation. That same horse who is now having a bit of fun at your expense had perhaps visited you in camp the evening before when they know they are safe, since your travelling is over for the day. You had petted him, fed him salt and rubbed fly dope on him, all of which he appreciated. But that appreciation was gone when it came time to saddle up.

I had a packhorse once that was full of these tricks, so I found a way to outsmart him. He was clumsy with hobbles but I kept putting them on till his fetlocks were real sore. The morning after I had let him go free, I walked slowly up to him voicing endearments and stooped down to remove the hobbles. Before he realized he had been tricked I had the halter on him. Strange to say I had no further trouble with him as long as I repeated the trick. Of course they are not all as dumb as that.

We started wending our way back to Entrance as Mr. Riviere wanted to send in his report and we were also getting low on supplies. I was expecting to meet Sam as we were now back in my district and I wondered what could have happened to him. No more work had been done on Eagles Nest Cabin and there were a few windfalls on the trail. I knew Sam was not lazy and could not understand it.

Pack train on the upper Sulphur River 1914.

FAY EXPEDITION, JASPER YELLOWHEAD MUSEUM AND ARCHIVES JYMA 84.87.65

The Bear Hunters

Mr. Riviere wanted to spend a few days at Rock Lake so I left him there and busied myself cutting firewood and clearing the trail. One evening when I was returning home I noticed a hunting camp over at the Wildhay River. Part of our duties as forest rangers included checking all travellers to make sure they were in possession of travel permits. This entitled them to the privilege of campfires and also enabled the Forest Service to have a check on their movements, because they had to state the route they intended to travel. Being ex-officio game guardians as well as fishery officers, we examined those licences as well. I found everything in order and initialled their permits.

The cook called me in to have a snack. During the course of our conversation he mentioned that one of the hunters was very sick and running a high temperature. I expressed concern at this and paid the invalid a visit. By this time his companions had returned from a walk up river and were feeling very downcast. I did my best to cheer them up and they told me they were from Iowa, men of moderate means who had been saving for years and pooling resources for this hunting trip.

There were four of them in the party and the one that I took to be the oldest wanted to shoot a bear in the worst way. It was a must, he said, as he had done so much bragging before he left Iowa that he would never live it down if he returned without one.

By this time it was quite late and pitch dark. As I slung my packsack on my back to leave, the bear hunter looked at me aghast. "My God, Ranger, surely you aren't going back to your camp tonight!" I said such was my intent and asked what was wrong with that. "Supposing you meet a bear, what would you do?" I said I supposed I would outrun him, besides I suggested, if he wanted a bear he would have to hunt him with a flashlight. "Not on your life," he said. I took my leave, saying I would be back in the morning to see how the sick man was doing.

When I told Mr. Riviere about the invalid he said the altitude was probably the cause of the illness. That made sense to me, as their camp was around 4,500 feet as compared to perhaps 200 to 300 feet at home in Iowa. I have also felt the effects of high altitude with ears drumming and trouble drawing a long breath. Riviere decided to take a hunt in the morning and try to get some fresh meat to tempt the invalid's appetite. As a rule, an outfitter will start out with only canned goods and a slab or two of bacon in the expectation of getting game upon reaching the hunting grounds.

Returning to the camp, I found no change in the man's condition and decided to move him to a cabin not far distant, which would be much more comfortable than a tent. The day was nice, we put him on a horse and two guides walked alongside and steadied him in the saddle. Riviere returned, having shot the heads off two blue grouse and the cook made some good soup. The sick man seemed to enjoy it, but his temperature remained the same.

We left for town the following day, as we could do no more for him. The party stayed a few days more, but as there was no improvement in his condition they decided that they must get him to a hospital and a doctor. He was getting so weak he could not help himself in any way. I don't know how many days they travelled as they had to make it in easy stages, but on the last day, when they were almost in sight of Entrance the sick man straightened up and asked for a cigarette. After resting for a few minutes he announced that although he was still weak he felt fine. They had to cancel their trip, though, as the altitude would have affected him again.

When we arrived in Entrance I decided to have lunch first and shave afterwards. I had not taken a razor along on the trip with Frenchy Riviere, so I had a dandy crop of alfalfa to take off. The little restaurant had changed hands and the new cook, Mrs. Holmes, was certainly wonderful and was soon to become famous for her pies. The railroad men were actually her best customers. They would never cook a meal in the caboose if they knew they were going to stop for water or be taking the sidetrack at Entrance. Mike Holmes, her husband, waited tables, looked after the poolroom and kept everything spick and span. When I presented myself for supper I noticed him eyeing me. "Didn't you have lunch here at noon?" he asked. I admitted that I had and he laughed and said that I looked like a different man without my whiskers.

I asked Mr. Badgley what had become of Sam Munson and heard a rather sad story. After Frenchy Riviere and I had left, Mr. Badgley and Ranger [R.E.] Hawkey[14] visited Rock Lake and found the cabin in a filthy mess. Sam had located a beaver colony and was using the place as an abattoir. Mr. Badgley made him get down on his hands and knees and scrub the floor and then get out and never set foot in the Forest Reserve again. I was sorry to hear this, as Sam in spite of his queer ways was really a fine fellow at heart. Years later I learned that poor Sam had been found dead in a lonely cabin, no doubt dreaming of pirate gold until the end.

[14] R.E. Hawkey was a forest ranger on the Athabasca Forest for five years from 1917 to 1921.

Forest supervisor Ward Badgley at Entrance Headquarters, around 1920.

ROY WOODLEY COLLECTION COURTESY OF MARY LUGER

North on the Lower Trail

After picking up our supplies, Riviere and I started out again. This time we travelled the Lower Trail, but I found it rather monotonous and missed the lovely scenery of the Mountain Trail. We saw very little game, as we travelled through timbered country most of the time.

A Montana Pete Story

At Big Berland River Cabin we met Ranger Milendorf, who made us welcome and we talked far into the night. He told us some stories about Montana Pete, who has been already mentioned.

It happened during the flu epidemic of 1918. The Mounted Police had been informed of the death of an Indian who had died by his campfire and, as it was winter, he had quickly frozen. It was necessary to hold a post-mortem and Sergeant Spriggs hired Montana Pete to act as guide, helper, and later as undertaker. They found the corpse with very little trouble, and after Spriggs had noted down the details, he told Pete to load the body on a packhorse.

Pete, besides being a powerful man, was also a great talker, and while he was lashing the remains on the horse, he remarked to Spriggs how fortunate it was that the body had been frozen in the shape that it did as it made packing that

Trapper and outfitter with trophy racks in the Brazeau Forest, 1940s.
ALBERTA FOREST HISTORY PHOTOGRAPHIC COLLECTION

much easier. When they arrived at the Forestry cabin at Little Berland River, they took the body inside in case a prowling wolf or coyote might mutilate it.

In the cabin, Pete kept on relating his experiences with the Mounted Police when he served as in interpreter in North Battleford. Finally, realizing that he was hungry he rolled up his shirtsleeves and said he guessed he would take a little wash and cook supper. "Good idea," said Spriggs, "and make it a darned good wash while you're at it, Pete." Pete was somewhat careless about his hygiene.

When supper was finished, both men grew drowsy as the cabin was hot and they had had a hard day. The corpse had been propped up against the wall by the door, which was also quite close to the stove. It started to thaw out and suddenly fell over sideways, knocking down a frypan which was hanging on the wall. One side of the face had thawed out also and what had formerly been a peaceful look had now changed to a diabolical grimace. This drew forth a torrent of oaths from Pete, which cannot be repeated here. He said the old fellow was a no good so-and -so when he was alive and now he was making a nuisance of himself when he was dead. "Anyway," remarked Spriggs, "he's in better shape now for putting in a coffin."

Pete was in a vile mood for the rest of the journey and did not get back to his own self until the last sod was placed and he had had a few drinks of bootleg whisky. Then everything was a big joke again as far as he was concerned.

Riviere and I continued on our journey, arriving at the Muskeg River Cabin, which was sixty miles from Entrance and the end of the telephone line. The Muskeg River had been wrongly named in my estimation as it was a lovely stream teaming with Dolly Varden trout and the water was clear as crystal whereas the creek that supplied the cabin with water bore a brownish tint. Nearly everyone agreed it was the poorest water in the whole Forest Reserve.

Riviere made a side trip upriver and I stayed at the cabin to do some baking and get in a spot of fishing. He hiked, as the cayuses were becoming trail weary and in need of rest. I had no trouble with them straying as they were knee-deep in grass in a large meadow. Riviere followed a trapper's trail along the river and stayed overnight at À La Pêche Lake where a Forestry cabin had recently been built. À La Pêche Creek emptied into the Muskeg River and, according to his story, the fish were so hostile in this creek that you had to hide behind a bush while you baited your hook. Years later I saw a movie taken by a member of a hunting party on that creek, and it was action all the way. The fisherman was landing beauties weighing three to four pounds regularly. Truly a fisherman's paradise.

Telephone Line Along the Old Trail – *the historic Aboriginal route north of Hinton, most of which was incorporated into the DFB's "Lower Trail".*

TELEPHONE LINE ALONG THE OLD TRAIL.
BY ROBERT GUEST, 1991. ACRYLIC ON CANVAS (ORIGINAL IS IN COLOUR)

Forest Rangers were called upon to do a variety of searches and rescues. Here they are returning to Entrance with the body of an unfortunate man from the north, c. 1920s.

ERNEST HARRISON COLLECTION, COURTESY OF LUCY MCRAE

Ewan Moberly's Homestead in Grande Cache, 1914.

FAY EXPEDITION, JASPER YELLOWHEAD MUSEUM AND ARCHIVES JYMA 84.87.67

Opposite page: Ewan and Madelleine Moberly and two of their ten children, Lactappe and Joe, at their home in Grande Cache, c. 1912. Ewan Moberly and his family were one of the six Métis families evicted from Jasper National Park in 1910 after it became a National Park in 1907. The fur coat and hat worn by Mr. Moberly may well have been the handiwork of Mrs. Moberly.

JASPER YELLOWHEAD MUSEUM AND ARCHIVES JYMA PA 20-29

The Grande Cache Settlement

Our next stop was Grande Cache. This was the Indian settlement that I mentioned earlier. There was a Forestry cabin here too, but it was poorly located as the cattle hung around it and messed it up badly. Nothing in the food line could be hung outside on account of Indian dogs and the water came from a well, which was not too reliable. I was told that when the hay had been all fed to the cattle and the snow was still on the ground, they were turned out to rustle. I was always under the impression that cloven-hoofed animals could not paw the snow away and uncover the grass underneath the way horses do, but these cattle must certainly have done it to survive.

There was a creek a few miles from here called Cowlick Creek, and there the cattle had found salt, sulphur and minerals, which they crave. They made good use of these licks in summer when the flies that attack their heels are so troublesome. They stand close together in mud and water up to their knees and, swishing their tails, keep the large horseflies away from each other.

The native women at Grande Cache were very skilled in needle- and beadwork. They had moccasins, buckskin coats, gloves and mitts, rifle scabbards and tobacco pouches all beautifully beaded. I bought several pairs of moccasins for winter wear and a fur cap that was the envy of many. The cap was a two-tone affair with the chocolate or almost black fur of the woodchuck making up the lower part and the silvery grey of the mountain whistler topping it off.

Years later I bought a lovely whistler or marmot robe from Mrs. Ewan Moberly who guides the destiny of the tribe.[15] Only the backs of the pelts were used and they were so perfectly matched that it was difficult to tell where one ended and the other began. It was Indian-tanned and sown with babiche. I had a taxidermist line it with felt and it was the equal or perhaps better than an

[15] Ewan Moberly had married Madeleine Findlay, descendant of David Thompson's guide, Jacco Findlay. Ewan and Madeleine were among the leaders in the Grande Cache community.

eiderdown robe. It puzzles me why this fur is not used more extensively, but there must be a reason I suppose.

The Indian fall meat hunt is quite an event, with all members of the family participating. The men do the actual killing and there their responsibility ends. The mighty hunter returns to his teepee and gives instructions to his woman and family to get the meat back to camp PDQ before an eagle, wolverine or bear beats them to it. If he has used more than one shell, he will tell the young fry to pick the others up, as they will come in handy for tipping the arrows used to stun rabbits.

Marmots are found only at very high altitudes and spend the greater part of their life in hibernation. They have their homes in rock crevices and will sit and watch the approach of an enemy until they figure the danger zone has been reached. Then they emit a shrill whistle and duck to safety. On one occasion I saw a little fox terrier almost driven frantic in a colony of them. He would charge one some distance away only to have it whistle and disappear. Then another would whistle nearby and he would try his luck there with the same result, and so it would go with nothing to show for all his efforts.

The Indian children generally have .22 calibre rifles and while mother is cutting up the meat they will shoot whistlers. They don't fool around either, for they know that the more they kill, the better their bed will be in the coming winter.

It is a busy time for the native women during this fall hunt. After the meat is brought to camp it is cut into very thin slices, then smoked and dried. If pemmican is desired berries have to be mixed with the meat after it is pounded.

The work does not end here, either. The women have to cut teepee poles and poles for their drying racks. The hides of the animals have to be smoked and

Babiche

Babiche is made from rawhide and has multiple uses. Hide is denuded of hair through a soaking process, stretched until dried and cut into long narrow strips. Named by the early French traders, these thongs were used for fishing and harpoon lines, lacing for snowshoes, bowstrings, gill nets, tumplines and headbands, lacrosse rackets and drumheads. Bags of fine workmanship were knit of babiche. Braided babiche was used in making halters and carrying straps. The hides of land and aquatic animals were made into babiche and varied, according to animal, in thinness, colour and strength.

Source: *The Canadian Encyclopedia*, online edition, 2001. http://thecanadianencyclopedia.com/.

Taking a break in the Smoky River Meadows near Grande Cache, c. 1920s.

HAZEL HART COLLECTION, HINTON LIBRARY AND ARCHIVES

tanned. Running amongst rocks raises hob with the moccasins, and patching and providing new footwear also falls to mother's lot. Truly, never a dull moment. But they seem to enjoy it and regard it as a picnic.

Riviere made a trip to the Big Smoky on foot and I experienced a little trouble with the horses, as the feed was not too good. I got around it by picketing two of them for a few hours and then changing them off for two others. The flies were bad so I fenced off a small pen and built a smudge, taking care not to let it blaze up. Whenever it showed signs of doing so, I threw on a little damp earth and rotten wood or green vegetation to damp it down. The horses certainly enjoyed the smoke and crowded in as close as possible. The building of smudges is something that the Indians very seldom do, so after awhile a bunch of their horses came charging in to smoke the pipe of peace. They were met with a very cool reception. A short, spirited battle took place in which no punches, or in this case kicks, were pulled.

Fire at Rock Lake

Although there remained a part of the Forest Reserve which we still had not covered, Riviere decided to wind up the trip as Grande Cache was the turning point. We decided to return on the Lower Trail, as the Mountain Trail would have added two more days travel and the horses were already weary. The return trip was uneventful, but when we reached Moberly Creek, we were sitting down to supper when the telephone rang. Mr. Badgley informed me that a Jasper Park warden had spotted a fire in the Rock Lake vicinity and he was rounding up a crew to go out. He told me he would also send out my packhorse as I had only used my saddle horse on the trip with Riviere. Mr. Badgley continued by reminding me to be careful and avoid getting myself or any of my crew cut off, as that would be a horrible way to die.

Dominion Forestry Branch Fire Rangers, Rocky Mountains Forest Reserve, early 1920s.

BOB STEVENSON, ALBERTA FOREST HISTORY PHOTOGRAPHIC COLLECTION

Dominion Forestry Branch Fire Ranger Badge.

ALBERTA FOREST HISTORY PHOTOGRAPHIC COLLECTION

Now, before I joined the Forest Reserve, I thought along the same lines as thousands of other Canadians regarding forest fires, fooling ourselves that we had timber to burn. When I pre-empted in British Columbia before the First World War, I had set fires indiscriminately to clear the land. Now that attitude was all changed and I had become fire conscious almost overnight.

I don't think that I slept any that night and was up, had breakfast, saddled up and was on my way before daylight. I parted with Riviere, who shook my hand and told me how much he had enjoyed my company. He said, too, that in a few years he might return, and if I was available we would explore the Jack Pine, Sheep Creek and Big Smoky districts, but I never saw him again.

I arrived at Rock Lake before the fire-fighting party turned up, so I left a note telling the ranger in charge to keep on the trail towards Eagles Nest, as I was almost certain that was where the fire was. I said I would try and have a suitable campsite located by the time they showed up.

Sure enough, the fire was burning in a flat that had been burnt many years before. There was a horrible mess of windfall there and actually the fire was doing more good than harm. A crew member later remarked that the windfall was so tough that a bird could not fly over it. I was fortunate in finding a small pool of water and decided to put our camp there. It was some distance from the trail and not too far from the Jasper Park boundary. This was one of the few places along the entire boundary that boasted a cut line through the green timber.

I got back out to the trail just in time to meet the firefighting party. Axes and crosscut saws were unpacked and we set to work cutting a makeshift trail into camp. After we unpacked, the horses were led back to the main trail and headed back towards Rock Lake where I was sure they would stay. Ranger Hawkey had

Len Langelle (Blackie)

Len "Blackie" Langelle was a long-time resident of Entrance until his death in the 1940s. He had a black beard and dark complexion, from which his nickname derived. Blackie had been a teamster on railway construction, driving multi-team horses and mules moving earth and rocks and grading before moving to "forestry" on seasonal trail and fire crews as Glen describes. He prided himself on caring for his horses and in later years looked after the AFS horses and barns at the Entrance headquarters. He had an improperly healed broken arm from being hit by a train, and used a yoke to carry water from the Athabasca River to his log cabin. He never learned to read or write, but did buy a daily paper to use as kindling to light his fire. First, however, he asked a neighbour to read the paper and tell him the major events. His greatest enjoyment was a radio given him by Harry Phillips. (Neil Gilliat *If Moose Could Talk,* 1998, Hart 1980) There is a L.Langill in the DFB list for 1913 and 1916 – this may have been Blackie.

Group picture at Woodley Bro's. general store in Entrance, early 1920s. Standing l-r are James Shand Harvey and Blackie Langelle.

MARK TRUXLER COLLECTION, COURTESY OF JACKIE HANINGTON

led the party of eight men out, but he had to head back the following morning in case another fire started elsewhere and he would have to pack out to it.

The mosquitoes were plentiful where we camped, but we had to put up with that discomfort for the protection that the waterhole afforded us in case the fire got out of hand.

I split the crew in two so we would have a night crew to protect the fireguard that was cut and dug during the day. I had the day crew myself and put an old woodsman who went by the nickname of Blackie in charge of the night crew.

Rumour had it that the name of Blackie had been hung on him because of his resemblance to a character in a serial running in the *Saturday Evening Post* at the time – Blackie Daw – a partner of a certain Get-Rich-Quick Wallingford, implicated in a number of shady schemes to obtain money fraudulently. It was a very amusing serial and Blackie Daw was described as possessing a dark handlebar moustache similar to the one our Blackie sported and was quite proud of. Blackie himself knew nothing of all this as he was quite illiterate.

So far we had succeeded in confining the fire in the windfall area and were gradually surrounding it with a fireguard. The object was to try to prevent it from gaining a foothold on a jackpine bench above, for if it did, nothing we could do would have stopped it.[16]

[16] The pine in this region is lodgepole, which is similar to the eastern jackpine with which it interbreeds in Alberta where their ranges meet. The term "jackpine" was often used generically to include both species.

One day it appeared as though we had lost the battle when a nasty wind sprang up. Sparks were starting fires in the tinder-dry moss and pine needles. Several backpack pumps had been sent out, which everyone on the crew regarded with disfavour, but they saved the day for us. They were made of heavy canvas with a pump attached and they were meant to be carried on the back, a sort of white man's burden, so to speak. The label attached gave the directions for filling them and the method of storing during the winter months to prevent mildew. It also portrayed a husky-looking individual operating one with a contented look on his face and a vast cloud of steam arising from the ground to reward his efforts.

The catch in the whole operation was that, providing you were successful in filling the pumps, the moment you mounted them on your back, the water started to seep through and run down your back, filling your shoes in its escape. One of the boys remarked later that he was sure he could have saved himself an extra trip if he had just wrung out his clothes and emptied his shoes!

Just the same they proved their worth that day. While one man sprayed water on a burning spot another would work with him, clearing away the duff and digging up the mineral soil which when mixed with water was effective in smothering the blaze. Everyone was almost exhausted when finally the wind died down and everything was under control again.

One evening we were lying around the campfire waiting for the cook to call us to supper when one of the young lads climbed up a nearby tree with an accordion, which he had brought along and began to serenade us. One of the men remarked that history certainly repeats itself. It was said that Nero fiddled while Rome burned and if accordions had been invented at the time he might have used one of them.

One day when the fire appeared quiet, I walked out to the Jasper Park boundary and noticed that the line was all cleared out and a line of hose was strung out with a pump at a water hole, but there was no sign of a crew anywhere. It is all right to be prepared, but if our fire had got away and swept in that direction the chances are they would have lost their hose.

As I mentioned before, fire prevention was in its infancy and we were having to learn the game the hard way. For instance, those pack pumps we were using have now been improved to the extent where, though still clumsy, they can be used without the operator getting soaked. Only one firm at that time had put forth a pumper unit and they were temperamental brutes that taxed one's patience to the limit. Sometimes you despaired of getting them started before the fire reached you and burned them up, and then again you would wish that it would. I had looked over the one and only pumper we then had and was anxious to try it out, but the stumbling block was the matter of getting sufficient hose transported to the scene of the fire. The hose was in rolls and in packing a horse the packs are always made as square or rectangular as possible so that the rope will stay on the corners while you throw the diamond hitch. This cannot be done with a round object.

Another ranger and I finally solved the problems by driving four stakes in the ground at the desired width and length and folding the hose inside, tying it in two places much the same way as hay is baled.

One evening it commenced to rain and then turned to snow. In the morning we had four or five inches. This finally killed our fire, but I wanted to be sure that it was completely out so I kept Blackie with me and sent the rest of the men home. It was anything but comfortable in the tent as we had no stove, but I did not mind it too much because it was my first forest fire and we had succeeded in keeping it within bounds.

I have often thought since that if we had had at the time one of the powerful high-pressure pumps now in use we could have put out the fire in two days instead of two weeks. Very little mopping up remained to be done. After the snow storm, Blackie and I found a number of logs burning underneath which we sawed off and turned over, shovelling snow on the hot spots. That did the trick and we headed back to Entrance.

The whole country seemed to be tilted on its edge. The felling of the timber was rather awkward on such steep places. One man would be down on his knees, while his partner would almost be standing on tiptoe.

Men and dog, Athabasca Forest. Extensive burns such as this were common but soon regenerated naturally to lodgepole pine, c. 1920s.

JACK GLEN PHOTOS,
ALBERTA FOREST HISTORY COLLECTION

CHAPTER THREE

ODD JOBS AND FAILED ENTERPRISES, 1921–1923

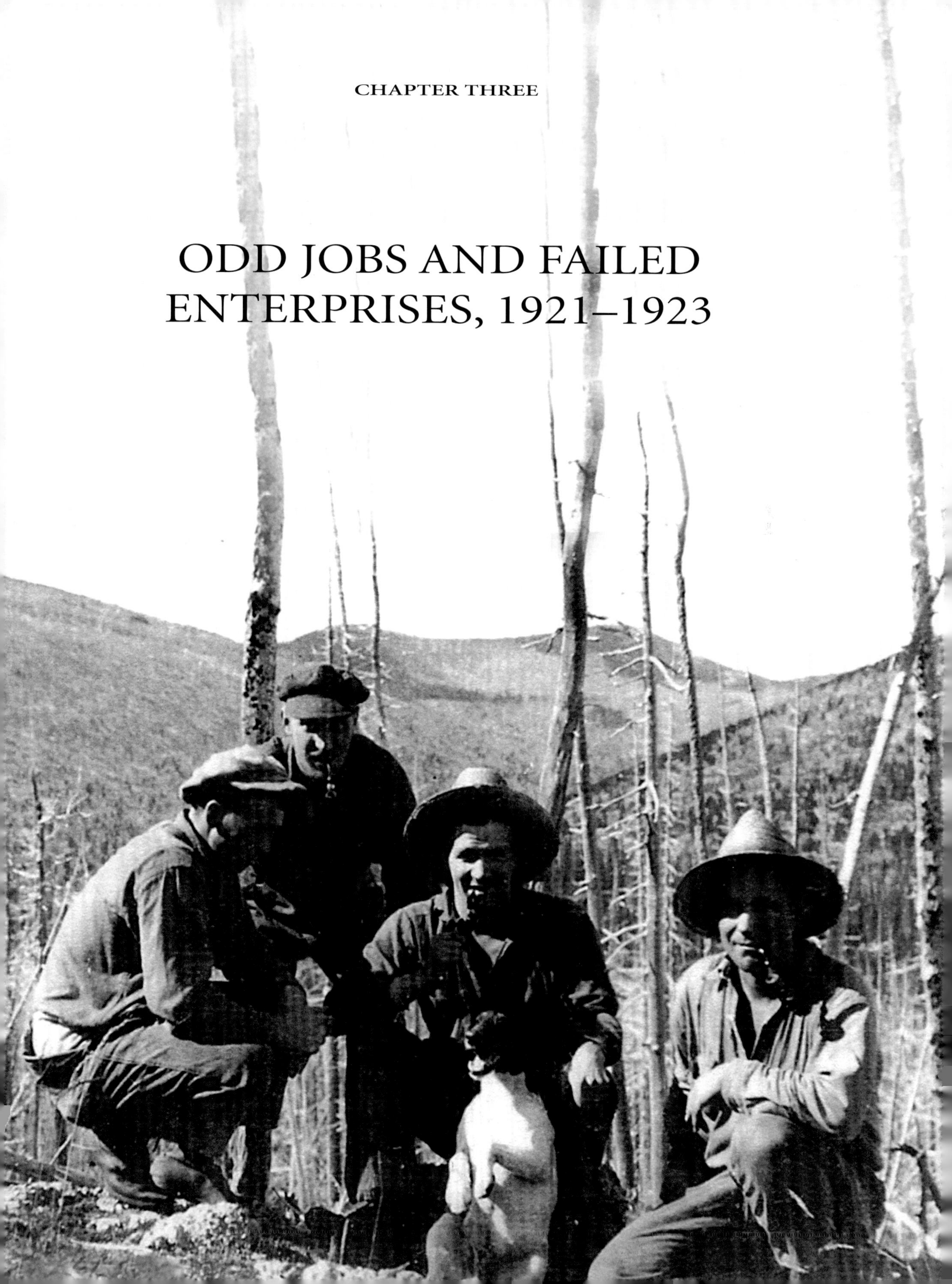

A Poor Farming Prospect

In the mail I found a letter from my sister in Edmonton telling me that her husband had decided to try his hand at farming and had acquired a farm at Calmar, Alberta, under the Soldier Settlement Board scheme. She asked me if I would be willing to come down and help them get started, and possibly enter into a partnership deal. Now my brother-in-law had had no previous experience in farming. When we were overseas I had rashly promised that I would help him get started as I had farmed prior to enlistment. Now I regretted my promise, as this roving life on the Forest Reserve had got into my blood. However a promise is a promise, so I tendered my resignation. Mr. Badgley said he very much regretted my decision, but hoped to see me back after I had initiated my brother-in-law into the farming game.

Jack Glen on the farm at Calmar, c. 1921.

JACK GLEN PHOTOS, ALBERTA FOREST HISTORY PHOTOGRAPHIC COLLECTION

Our first season on the farm (1921) showed a deficit. The farm that he had acquired was in a very run-down condition so we decided to summer fallow it. This would allow the land to recuperate and many of the weeds would be killed. In the meantime we had leased a hay farm adjoining us, which normally should have been a paying proposition as baled hay was in demand. Prices were high, because at that time machinery had not yet taken the place of horses to a great extent. Unfortunately we experienced a dry season with the result that we only harvested about one-third of the crop that normally we should have.

Early that same season, a neighbour of ours who had been clearing land around the edge of a marsh the previous fall had one of his fires come to life and start what might have been a disastrous prairie and bush fire. When it started to spread he telephoned his neighbours asking for help. The response was very gratifying and soon we had a number of willing helpers. Each had brought along barrels of water and shovels.

I was asked to take charge of operations, as I happened to be the only one with any firefighting experience. Fortunately this farmer had a ploughed field that ran almost the length of his place on one side of his farm. On the opposite side a road allowance joined another road allowance at right angles so that we had a perfect fire guard in the ploughed field, a fair guard in one road allowance, and a not too good guard in the other road allowance, which had just recently been partly completed and was not yet gravelled.

If the fire jumped this road my brother-in-law's place would be right in its path. I decided to fight fire with fire and directed the crew to start fires along the edge of the road, reminding them to be very careful to put out any sparks that jumped over. Well, I suppose some of these people figured that I had taken leave of my senses and one or two of them said they had come to put out fires, not to light them. I told them I had no time to argue with them and to do as they were told and I would take the consequences.

The fire was now gaining in volume every minute but our backfire was beginning to reach out to meet it and none too soon, for the heat and smoke were fast becoming unbearable. I then took a chance and told the people to back up and get some fresh air. They should however be ready to put out any fires that might be smouldering in the dry peaty soil of the road allowance. We stayed long enough to be certain that all fires were out, then went home very tired, but thankful that the fire had not got away.

Well, if our first year was poor, our second year (1922) was worse. The well ran dry and a new one had to be drilled. The creeks ran dry. The crops dried out and what grain there was, was hailed out. That pretty well spelled our finish. My savings were gone and though there might be a bare living in the farm for one, there was certainly not for two. I left the farm and to tell the truth I was not sorry, for I had longed for the mountains during those last two years.

Back to the Rocky Mountains Forest Reserve – Fire on the Berland

When I made a trip to Entrance that fall I found quite a few changes. Mr. Badgley was gone. The brothers had dissolved partnership in the general store. The younger brother, Earl, had opened a store at Red Pass Junction in British Columbia. There were several changes in ranger staff, but Louie Holm was still there and he took me over to meet Mr. Badgley's successor, Tom Burrows. The latter was pleased to see me, though he explained that they couldn't sign me on as a ranger that late in the season but they would be glad to employ me at patrolman's wages if I would care to go out and look after a fire in Berland River District. They would supply me government horses, as mine remained behind at my brother-

Dominion Forest Rangers, Athabasca Forest Reserve, c. 1922. L-r., Tom Monaghan (store owner), J.E. Cook, Ernie Harrison, Tom Burrows (forest supervisor), T. N. Irwin.

ROY WOODLEY COLLECTION, COURTESY OF MARY LUGER

Dominion Forestry Branch 'Forest Officer' badge.

BOB STEVENSON, ALBERTA FOREST HISTORY PHOTOGRAPHIC COLLECTION

in-law's place in Calmar. The fire did not warrant sending a crew as it was in an isolated spot in the mountains and rain or snow should not be too long coming. I gladly agreed to go and it was arranged that a younger ranger by the name of Jim Mills[17] would go with me.

[17] Jim Mills was a forest ranger on the Athabasca Forest in 1921 and 1922.

FOOTNOTE SOURCE: WILLMORE WILDERNESS PARK MAP, 2007. ALBERTA TOURISM, PARKS, RECREATION AND CULTURE AND THE GRANDE CACHE HISTORICAL SOCIETY.

That turned out to be a real hoodoo trip. The first day we reached Moberly Creek quite late. Next morning we rose early but not a tinkle of a horse bell was to be heard. Mills figured they might be lying down and said that if I would get breakfast he would go and get them. Time went by and there was no sign of him so I cut myself a willow pole and went fishing for rainbow trout. I did not go very far from the cabin, of course, as I expected him back any minute. By noon I was really worried. I cooked and ate some lunch and had just finished when he showed up, hungry, wet and disgusted. He had followed the horses for miles down the creek and the Wildhay River, and had fallen off a log when he crossed the creek.

I saddled and packed while he ate and changed into dry clothes. We left Moberly about 2:30 that afternoon expecting to reach the fire that evening. Bad luck still seemed to dog us. It was all right until we left the Lower Trail and headed up the Big Berland River where there was practically no trail. The packs banged against trees and got knocked off balance and had to be straightened time and time again. Darkness overtook us but we still kept travelling. I did not like it a little, either, as I always have a dread of having my eyes poked out with the limb of a tree.

Grave of Pierre Delorme at Big Grave Flats, Fay Expedition, 1914. Pierre Delorme was a blind, 6 foot 8 inch (203 cm) Aboriginal man. While riding through Rocky Pass in 1906 he fell off his horse and hit his head on a rock. He died soon after and was buried in this meadow at the base of Sheep Mountain along the Sulphur River, the large "spirit house" serving as his grave site.

JASPER YELLOWHEAD MUSEUM AND ARCHIVES JYMA 84.87.53

It must have been around midnight when we finally noticed the glow of the fire. We had now left the river and were in rolling country, so we decided that when we next heard water running (it was too dark to see it), we would make camp regardless. We finally heard water and stopped. Jim had to climax it by stepping into the creek when he dismounted. I found a few dry twigs, which I managed to coax into a flame to give us light enough to show us some more wood and by the time Jim had unpacked I had a rousing fire going.

Supper over, we lost no time going to bed. I saw a sort of fence just within range of the firelight and assumed that some prospector had built a smudge for his horses as I did at Grande Cache. Now by tying a canvas to the fence I could have a sort of lean-to tent, which would keep the white frost off me. It had been real frosty at night of late. I was soon in bed and asleep, but Jim decided to sleep as close as possible to the fire as he was wet and cold. He drained the water out of his boots and put them close to the fire hoping they would be dry by morning. We left the horses tied to trees, figuring they would not suffer too much until daylight, when we could find food for them.

A ptarmigan call of "wick, wick, wick, wick" woke me the next morning. The vegetation was white with frost but the sun was showing over the mountaintops and gave promise of another fine day. Then I got quite a shock. What I had figured to be a smudge pen turned out to be an Indian grave. The customary roof was missing and in its place a few rocks were mounded up, with a wooden cross, badly rotted, lying on top. I hate to think how I would have felt if I had wakened and seen a grinning skull like the one I mentioned before at that other grave, lying about two feet from me. I lost no time in getting my bedding away from there. The corpse however had other neighbours, permanent ones. A huge anthill was on the other side of the grave.

Down by the spot where the campfire had burned itself out Jim was still sound asleep with his head covered over in his sleeping bag. I shook him awake and then noticed his boots, or rather what had once been his boots. They had been burned almost to a cinder. A small juniper bush had caught fire and the fire had crept along the root where the boots were standing. It was no laughing matter, but I could not contain a chuckle or two when I looked at those mummified objects that had adorned a storekeeper's shelves a short time ago. Jim even laughed at himself after he had got over the first shock.

"Now look here, my flannel-footed friend," I said to him, "you are now elected Chief Cook and Bottle Washer and I will go forth into the wilderness and smite the fire alone." I led the horses to a patch of grass, hobbled them and in the meantime Jim had breakfast cooked. When we were eating I remembered I could perhaps help my companion out of his dilemma. I had bought a pair of Indian moccasins at the store in Entrance thinking they would be comfortable to wear in the evenings. When I got them out of the pack Jim said they were just like manna from heaven.

The fire appeared to be burning some distance up the creek, so I took a shovel, a mattock and canvas water bucket and started off. It had done very little damage so far, although it had spread over about three acres. However, it was rocky ground and almost at timberline, so it had very little fuel to encourage it. When it reached a clump of juniper it blazed up fiercely and threw up a dense cloud of acrid smoke. It was now in a deep ravine through which the creek flowed that we were camped by, so I figured that if I could prevent it from crossing the creek it would soon burn itself out. I hurried back to camp for lunch, and told Jim I would probably be out quite late working on the fire. It all worked out as I hoped and I had very little trouble holding it.

You may be sure I picked a new spot for my bed that night, and I advised Jim not to sleep too close to the fire in case he lost his shirt this time, and I would be hanged before I'd share mine with him. Next morning I climbed up to the burn and could see no danger spots anywhere so we packed up and started off for Little Berland Cabin. We parted company there, Jim heading back into his own district.

Back on Patrol

Tom Burrows had told me I could finish out the balance of the fire season travelling around and getting to know the country. He said also that a trail had to be cut down the Little Berland River, and from its junction with the Big Berland River on to the 15th Baseline, and I might look this over.

This was what I decided to do first. I had reached a large open meadow named Donald Flats[18] when I noticed a herd of Hereford cattle. Nobody had ever mentioned cattle to me so naturally I was curious. I finally saw a camp and rode over to it. A tall, elderly Frenchman, Joe Biznet by name, came out of the tent and asked me if I would join him in a cup of coffee. I accepted his kind offer and in the course of conversation asked him who owned the cattle.

It appeared that a rancher by the name of Scott had trailed them down from the Peace River country, bringing also a team of horses and wagon loaded with a mower and hay rake. This must have been a nightmarish trip as there was practically no trail, and even the way I had travelled that day was over a trapper's trail only meant for winter. I remembered then that I had noticed wheel tracks at the river crossings but had paid little heed to them. It reminded me of some of the stories I had read about the early pioneers who had crossed the continent in covered wagons and settled the West. The only difference between Mr. Scott and the early settlers was that he did not have to fight off hostile Indians.

Biznet and Scott had cut and stacked some hay and Biznet had agreed to stay there and feed the cattle that coming winter. The old man was lonely and asked me to stay the night, which I did.

He told me about the wonderful dog that Scott had, saying that if it had not been for that dog he doubted that they would have made the trip. I thought he

[18] Donald Flats is named after Donald McDonald, a Métis from the Red River area who came west with the troops at the time of the Riel Rebellion. He married widow Louise (Findlay) Thappe, descendant of David Thompson's guide Jacco Findlay. McDonald trapped and guided extensively throughout the Athabasca Forest area, and several other places are named after him. Source: Susan Feddema-Leonard, *The People & Peaks of Willmore Wilderness Park*, Willmore Wilderness Foundation, 2007.

Cutting trail through old burns is a difficult proposition. Supervisor Stan Clark on Athabasca Forest, 1915.

DOMINION FORESTRY BRANCH, ALBERTA FOREST HISTORY PHOTOGRAPHIC COLLECTION

was exaggerating but later I saw for myself just what an intelligent animal Lassie really was. She was a border collie and I am not sure whether Scott trained her himself or not, but whoever did deserved great credit. The following year (1923), Joe Biznet was doing some work for the Forestry with his team. He had turned his horses loose to graze the evening before, with bells on of course, and after breakfast he called Lassie and told her to go and fetch them. I could not hear even the faintest sound of a bell, and as time went by I fully expected to see the dog return without the team. I had misjudged her, for like Little Bo Peep's sheep they came home wagging their tails behind them and on high, too, with Lassie taking care that they did not loiter and ready to nip their heels if they did.

Biznet told me he was figuring on shooting a moose, as they did not want to butcher any of the cattle, since they were not in too good shape after their hard trip. On a foggy morning some time later, Joe saw what appeared to be a moose across the meadow from their camp. He picked up his rifle and shot. The would-be moose dropped, but when he reached the spot he found to his sorrow that he had killed his own horse.

Emma Nickerson's New Truck

Later on that trip I met a hunting party returning to Entrance. The outfitter was a chap by the name of Herman ("Nick") Nickerson whom I had met a number of years ago at Burns Lake, B.C. At that time, the Grand Trunk Pacific Railroad was being built to Prince Rupert, and he was a telegraph operator. A handsome girl was with the hunting party as a horse wrangler. Her name was Emma Thorson. She later married Nick Nickerson. She was Norwegian and very powerful, and no work was too hard for her. Later [in the 1950s] she bought a gravel truck and hauled load for load with the other truckers in building the Jasper Highway.

Every year the Indians go on a pilgrimage to Lac Ste. Anne. This is quite a long journey, even by train, as the trains have to change and sometimes the connections are not too good. Hearing that Emma had bought a truck, the Indians asked her if she was willing to transport them. She agreed and to simplify the loading of the baggage she took out the tailgate. She crowded as many as possible into the truck and prepared to set out. She had driven the truck very little yet, but it took more than that to worry Emma. She started slowly and then by mistake she pulled the wrong lever – the dump lever – and up shot the box and deposited the whole caboodle on the road. "Well," she said, "now we know that the dump works perfectly." The Indians thought it was a great joke and I understand that the kids wanted a repeat performance. There were no further incidents on the road as far as I could learn. Emma should really have been a man instead of a woman as she hated housekeeping and much preferred the wide-open spaces.

A Short Career in Logging and Sawmilling, Winter 1922–23

My season as patrolman drew to a close, and I prepared to leave Entrance for the winter. I was in a much happier mood than when I left two years before, as Tom Burrows had told me to report for duty on the Athabasca Forest Reserve the following spring (1923). Ranger Jim Mills had decided to quit, so I bought his eiderdown and other odds and ends. We travelled to Edmonton together and after a few days in the city, he left for Oregon and I never saw him again.

One day when I was in the employment office, Mr. Hadley, a sawmill operator, came in looking for a small crew of experienced bushmen. The wages were deplorable, but I decided it would be as good a way as any to put in the time, and I went to the millsite at Styal. It was only a flag station and the people around were all engaged in farming and logging. The nearest town was Evansburg and across the river was Entwistle. These were both coal mining towns and their occasional dances did much to relieve the monotony of camp life.

Logging camps in those days were a far cry from the comfortable sanitary camps of today. Each man carried his own blankets and slept in bunks of hay or spruce boughs.

Emma Nickerson

Emma Nickerson was a colourful and remarkable character in the Hinton area for many years. As Emma Thorson she married Herman Nickerson, a telegraph operator at Entrance. By the late 1920s they worked together guiding hunting parties for extended trips. Emma had her licence as a guide and outfitter, and they worked their business for almost 30 years until they entered the trucking and gravel business around 1950. Their gravel pit near Prairie Creek was a good source of gravel for North Western Pulp and Power when it started operations in 1955. Nickerson Creek is likely named after them. (Hart, 1980)

Emma Nickerson, c. 1920s.

CAIRNS, HAZEL HART COLLECTION. HINTON LIBRARY AND ARCHIVES

Typical portable sawmill, Brazeau Forest 1927.

ALBERTA FOREST HISTORY PHOTOGRAPHIC COLLECTION

I was often the handyman around the mill, but one day I was working the cut-off saw. This saw fitted into a recess in the back of the bench and if any lumber had to be trimmed it was slid along the bench and cut to the proper length. To do this trimming you pulled the saw out by its handle, zipped off the offending end and released it and it shot back out of harm's way. The power for the saws was supplied by a huge wood and coal burning steam tractor, but generally wood when we were sawing. The wood for the firebox was cut in suitable lengths by the operator of the cut-off saw and that was what I was doing that evening by the light of a smoky kerosene lantern.

It was hard work and one had to work fast to keep the bench clear. I had just cut off a length and released the saw when a large piece of bark, unseen by me, flew in and wedged in the recess. When the saw hit it, it bounced right out again and hit me on the hand. I was wearing double mitts, leather over woollen ones, and the leather one was ripped to shreds. The woollen one was nearly torn off and I almost passed out with shock and pain, as though lashed with a whip. It took me awhile to muster up courage enough to pull off the remnants of the mitts as I was sure that I had a bunch of sawed-off fingers inside them. The woollen mitt was just hanging by a few threads but there was not a scratch or drop of blood.

That was certainly a relief but I was firmly resolved to have nothing more to do with a cut-off saw. I went to the office and told the boss what had happened and that I was quitting. He said he did not blame me for quitting under the circumstances, but said that if I wanted to change jobs he needed two fallers and if I wanted to take it on I could choose a partner from the crew. I took him up on it, but for a long time I shuddered whenever I heard the whine of a saw.

I had been chumming with one of the fellows who had signed on in Edmonton the same time as I did, Bill Douglas,[19] a well-educated Scotsman. Bill was a wanderer. He had served as an instrument man with a survey party, then taken a homestead in the Rolla district of Alberta, and when he tired of that he just walked off and left it. Well, Bill was the man I chose for a partner and the two of us felled timber until the mill was shut down in order to make some alterations. Rather than sit around and pay board, we drew our time and went to Edmonton to enjoy city life for a few days.

[19] Bill Douglas was Glen's logging partner in 1920-21 and became a forest ranger on the Athabasca Forest from 1923 to 1929.

Tie Camp on the Embarras River

We were not too well fixed financially, so we soon got back to work, this time at a tie camp on the Embarras River, a tributary of the McLeod River. To get there, we got off the train at the flag station of Bickerdike and travelled from there by team and sleighs.

The camp was quite large, the food was excellent, but the living conditions were horrible. The bunkhouses were long rambling log buildings, housing fifty men each, with roofs of poles covered with spruce boughs. This was all right providing there was a good covering of snow and the weather was cold, but when a Chinook wind started, the snow would melt and run through, bringing with it spruce needles and misery to the inmates. The bunks were arranged in two tiers and were referred to as Muzzle Loaders because you dived into them head first from the end and reversed the procedure getting out, backing out feet first.

For light you had a Coleman lantern at each end of the building and for heat a huge barrel-type stove known as a Camp Comfort, which was also supposed to dry your wet clothes – quite an impossibility. I noticed several of the men

Coal Creek logging camp, Highwood River, Bow River Forest 1911. These log buildings are typical of early bush camps and appear very similar to Jack Glen's description of the logging camp on the Embarras River.

ALBERTA FOREST HISTORY PHOTOGRAPHIC COLLECTION

Hewing a square timber smooth with an adze. Railway ties were initially hacked with a broad axe before sawmills were introduced, sometimes finished with an adze in this manner.

DOMINION FORESTRY BRANCH, ALBERTA FOREST HISTORY PHOTOGRAPHIC COLLECTION

scratching vigorously and had a strong suspicion they were harbouring cooties.

The lights were put out every night, except Saturday, at nine o'clock. Getting to sleep as soon as the lights were extinguished was a must as far as I was concerned because of the ensuing "babel" that always began soon after. A teamster would be talking to his horses, accompanied by a number of characters snoring in several different languages. One night I was awakened by a poor unfortunate soul who had got out of bed to stoke the fire. The night was cold and no doubt he was not overburdened with blankets. Before I could get back to sleep a budding Caruso burst forth with a rendition of Hiawatha's Melody, ending abruptly in a violent fit of coughing, perhaps as an encore. You might occasionally hear a muttering of Robert Service's works, the most popular being the exploits of a certain Dangerous Dan McGrew.

All this was bad enough, but those bunkhouses were regular fire traps. If a fire had started, what possible chance did fifty men have to get through one door? It would have been a funeral pyre for at least some of the occupants. I worried over this eventuality until I hit on the solution. Why not evacuate through the roof? I had tried the roof poles and found that they moved without too much effort, so with this thought in mind I occupied an upper bunk. Bill and I made up our minds that just as soon as we had earned enough cash for a travelling stake, we would quit.

As I mentioned before, this was a tie camp and most of the ties were hewn. There was a limit to the size of tree a man could hew with a broad axe, and the company had a sawmill to saw the oversize trees. They could saw square ties out of sixteen-foot logs and get a quantity of side lumber besides. We cut these logs on a contract basis and soon found that we could make good wages this way. We could work when we pleased and did not go out when it stormed. This gave us a chance to get caught up on the sleep we might have lost at nights, and we could shave and do our washing. On Sundays we would work to make the price of our board and a little besides for the proverbial rainy day.

When we finally figured we had made enough, we were issued with a time slip to be cashed in Edmonton with the company that had hired the tie camp people. We experienced some trouble doing this in the city, however. The company was not dishonest, but not enough ties had been delivered to the siding and they were just being cautious. I assured them that everything was all right and that there was a huge stockpile of ties back at camp, but due to the condition of the river ice, sleighing had been stopped. This seemed to satisfy them and we were given our cash.

Our next job was at Shere on the Fraser River in British Columbia. Here we found conditions more to our liking. It was a new camp and had an excellent cook. The bunkhouses housed six men each and they were clean and comfortable. One thing I noticed about this camp – nobody was ever refused a meal. Hoboes from the freight trains got one meal and if they wanted to work they could stay as long as they pleased, providing of course that they were any good.

Bill and I went falling as usual. We crossed the Fraser River on the ice and had a stiff climb of almost a mile on the other side. The whole country seemed to be tilted on its edge. The felling of the timber was rather awkward on such steep places. One man would be down on his knees, while his partner would almost be standing on tiptoe. I told Bill that this would be good practice for him if he was going to get a Forestry job, as Entrance was built on a side hill. One old-timer told me that after a generation or two, dogs were born with legs longer on one side than the other. He said it was kind of awkward for them when they wanted to go back the way they had come, but that they got used to it!

Robert W. Service (1874-1958)

A bunch of the boys were whooping it up
    in the Malamute saloon;
The kid that handles the music-box
    was hitting a jag-time tune;
Back of the bar, in a solo game,
    sat Dangerous Dan McGrew,
And watching his luck was his light-o'-love,
    the lady that's known as Lou.

Excerpt from Robert W. Service's "The Shooting of Dan McGrew," *Songs of a Sourdough*, 1907.

There were five gangs of fallers besides ourselves. We took our saws back to camp every evening and the saw filer had a sharp saw for us each morning. After we had worked there for about a month, the woods boss said he had to reduce the falling crew by two gangs, as enough logs were on hand to warrant starting up his mill. He left it to the fallers to decide who would move to the mill. I asked what jobs were vacant and he said the saw filer was quitting. I could have that job if I wanted it, but it could be done in half a day and I would also have to serve as handyman for the rest of each day. I took the job but Bill decided to stay with falling and took a partner from one of the other gangs.

I enjoyed the new work very much and would have been happy to stay with it, but I knew the mountains would keep beckoning and anyway my future was secure with the Forestry. The blacksmith was very skilful and I helped him occasionally and learned a lot, especially about horseshoeing, which stood me in good stead in later years, as we had to shoe our own horses in the Forestry. I sometimes helped the millwright, but that was generally on Sundays and I was paid overtime. Occasionally we worked after supper as a number of mechanical bugs had to be found and corrected, as is usual in a new mill.

A comical incident happened there. The logs, so far, had been pushed onto the jackladder from the river ice and so up to the saw, but for the summer work, a pond would be necessary with piles driven and boom sticks strung around to hold the logs that would be driven down the river. A pile driver was rigged up but it was found impossible to start the piles because the gravel was frozen. It was decided to blast holes below the frost.

The company employed a number of Chinese men who handled the lumber as it came from the saw. Scotty Galbraith, who supervised the loading of the lumber wagons, knew that the blasting was soon to take place and when the cry of "FIRE" was heard, he signalled to the Chinese to take cover. Their interpreter was absent, and when they saw that everyone else had quit working they figured on taking a rest also. Out came their brown rice straw cigarette papers and they commenced rolling their own. Scotty made a last desperate gesture to them, but was greeted with deadpan expressions. Muttering something not too complimentary about Orientals in general, he sought protection under a loaded lumber wagon.

The blasting began and the gravel flew everywhere. So did the men. Luckily nobody was hurt. Not long after this incident, Scotty ducked under a wagon to fix a wrapper, and every last man in camp ducked for cover.

The District Forest Ranger visited the camp periodically and frowned on the existing fire hazard in the camp area, warning the operator that unless something was done about it he would be forced to close the camp down. An extensive alteration in the mill was planned and during the time required to complete it there would be no lumber for the Chinese to handle. This was a good chance to pile and burn the brush. They were instructed to take down all standing trees with the idea of pulling the stumps later and making a piling ground for lumber. The white men who worked in the mill were given other jobs in the bush and of course the fallers worked as usual. Naturally the Chinese required tools for their work. Their interpreter would fetch one man at a time over to me for tools. I had resurrected every antique saw I could find in the camp and he would go through much the same ceremony as a waiter reciting the bill of fare, his approach always the same. Smiling broadly he would bow to me, "Plese mista you giff saw this man." I would then pretend to minutely examine an old wreck and hand it to him, and he in turn would hand it to his man. Then he would turn to me and thank me profusely, giving another elaborate bow and backing away for a few steps before moving off. During that period, I was undoubtedly the most popular man in camp, in the Chinese eyes at any rate.

At last they were all equipped and then the fun commenced. They would saw or chop at a tree until it started to crack and fall. Everyone would drop his tools, put his hands above his head and flee, chattering, in all directions. This was probably their version of the lumberjack's warning cry of "Timber."

The End of My Logging Career

We were well into April of 1923 and that meant I would soon have to report for work at Entrance. I told Mr. McDougall, one of the partners at the mill, that I was quitting. He offered Bill and me a proposition for a logging contract, which seemed like a good thing, and had I not had the Forestry job to go to I might have seriously considered it. They had a tract of lovely spruce on a bench above the Little Shushwap River and they wanted the logs cut and dumped in the river. They would take over from there. We could use whatever method we wanted, but he recommended the use of chutes. I could see this would be the answer, providing the incline was not too steep, in which case the logs would gain too much momentum and jump the chute, ending up somewhere where they would be hard to salvage. He offered to outfit us with the equipment required, promised we would not lose money, and was sure that we could make better than wages. I asked Bill if he would consider it on his own, but he said no, not with anyone but me. We delivered our ultimatum to Mr. McDougall, who said he was sorry, but that was our own business.

I paid a visit to Edmonton to purchase an army surplus packsaddle that was cheap enough that I bought two. Horses were now much cheaper at Entrance than they had been two years before.

Bill Douglas had also applied for work in the Forestry, and I supported his application. He decided to stay at camp until he heard if he would be accepted, but soon heard that he was hired as an assistant forest ranger in the Athabasca Forest Reserve. He stayed with the Forest Service for about seven years, then got the itchy foot again.

The crew I had that year were all green men except for Blackie, whom I had worked with in my first year with the Service.

Forestry camp at Mount Bess, Athabasca Forest, 1912.

DOMINION FORESTRY BRANCH, ALBERTA FOREST HISTORY PHOTOGRAPHIC COLLECTION

CHAPTER FOUR

BACK ON THE TRAIL, 1923–1924

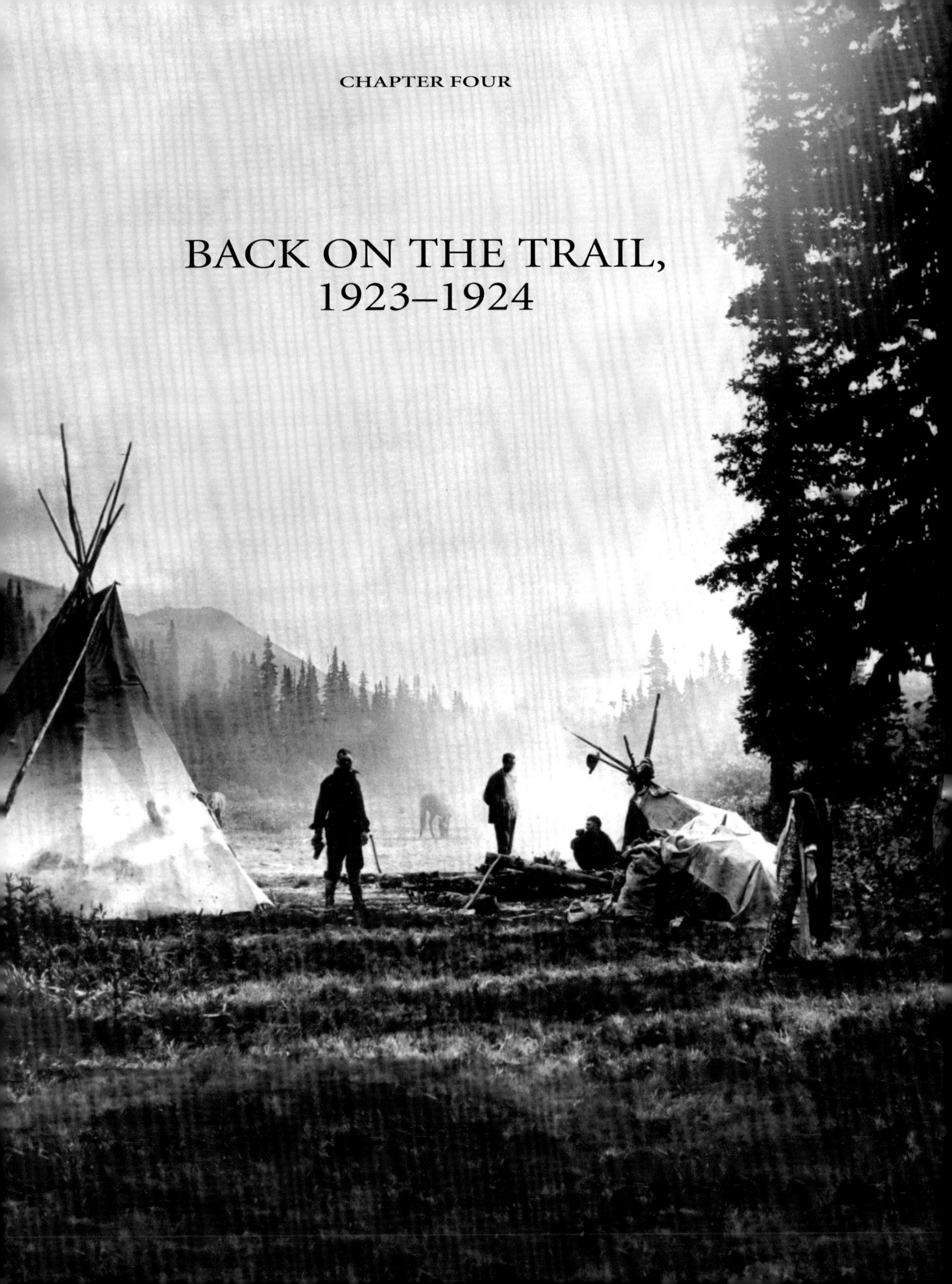

I returned to the mountains and more work on the trails that I knew so well. When I arrived back at Entrance in the spring of 1923, Tom Burrows told me that he wanted me to take a trail crew that season, but that it would not be in mountainous country. Ernie Harrison[20] had the Mountain Trail for that season and as an assistant he had a young German lad with a morose manner who, the following winter, committed suicide while on his trapline.

[20] Ernie Harrison was a long-time forest ranger on the Athabasca Forest and resident of Entrance. He worked with DFB in 1913 and 1917 then from 1922 to the fall of 1939. He also worked as a packer and logger.

The Bronc Buster

One morning in the Forestry yard, all was hustle and bustle with the rangers getting ready to leave for their districts. I was going out to Wildhay River Cabin to fetch in the horses that we needed. The horses that Jim Mills had owned had been purchased for Forest Service use and Mrs. Burrows, who was a keen horsewoman, had been riding one of them. That horse was a stupid sort of brute, but he had one thing in his favour – he could be caught any time and any place without trouble.

Apart from that, he was a barnyard pest. He seemed to have the knack of finding any open door or open gate. If feed or hay was being hauled down to the barn, he would tag along on every trip, picking any wisps that dropped off the load. He was also a lone-wolf sort of horse and this characteristic nearly caused his death a year later. We had him on our pack train and every time we moved camp he was sure to be missing. This went on all summer and on the morning we were starting to head home, he was not to be found anywhere. Everyone set out in different directions to look for him, but eventually we could not delay any longer and went without him.

The following spring two of the boys passing our old camp noticed fresh horse signs. One of their horses picked up his ears and whinnied. He was answered immediately and out of the bush appeared the horse we had given up for dead. He was a living skeleton and I still can't figure out how he survived the winter, as that was a deep-snow belt. The boys said that he was so glad to see the other horses that the tears were running down his cheeks. I think that was a slight exaggeration, but anyhow he learned his lesson and never again left the other horses.

That morning in the Forestry yard, I decided I would take that Mills horse and leave my usual mount for Mrs. Burrows to ride. That Mills horse had stepped on her foot one time when she was saddling him, and anyway he was a halter-puller.

My saddle was the Association type, used by the competitors at stampedes, and had a flank cinch besides the ordinary cinch. When this flank cinch is tightened, the fireworks usually start as the broncos heartily detest it. I liked that type of saddle and, although they cost more than ordinary saddles, they pay for it in comfort.

I had forgotten about this horse being a halter-puller and when I had the front cinch tightened up I gave the flank cinch a good hefty jerk and everything happened at once. He put back his ears, lunged back, broke the halter rope and

Jack Glen astride his favourite horse, c. 1940.

JACK GLEN PHOTOS, ALBERTA FOREST HISTORY PHOTOGRAPHIC COLLECTION

Association saddle.
BOB STEVENSON

The Association Saddle

The popularity of the Association Saddle continues with riders today. This saddle was custom made by Les Ertman of Onoway, a saddlemaker since 1962, and presented to Assistant Deputy Minister Cliff Henderson of Alberta Sustainable Resource Development at his retirement dinner March 8, 2008. The saddlebags were a gift from the Foothills Model Forest, where ADM Henderson served as a Board Member.

tore around the yard putting on as fine an exhibition of bucking as you would see at any stampede.

The boys loved to watch bucking broncos and every last one of them was grinning from ear to ear, hoping to have some fun at my expense. They quickly found themselves ringside seats on the top rails of the corral. I could not change horses without losing face, so I had to put on a show for them whether I wanted to or not. It seemed like a hot day coming up and I had decided to make the trip without chaps, but now I dug them up and put them on, likewise my spurs, as they are a must when riding a bucker. The spectators started to kid me, asking what sort of flowers I preferred and giving advice. One suggested that he would send for the Prince of Wales – who at that time had the reputation of being a somewhat indifferent horseman – to come and take the rough spots off this old gravedigger.

Now I had a trick up my sleeve which I thought might work, so after getting a new piece of rope and a crouper, such as is used in some light driving harness, I caught up my bronco. He had now quit bucking but he was still snorting and looking wild eyed. I led him back to the hitch rail and passed the rope through the halter ring, back to the crouper under his tail and then tied it to the hitch rail. Going in front of him I slapped him on the face with my hat. He reared back but as soon as the pull came on his tail he shot ahead and stood trembling. It is claimed that a pull on a horse's tail affects the nerve centre. I saw a horse once with a broken tail and his hindquarters were paralyzed. He had to be destroyed.

Well, I figured that most of the fight had gone out of my bronco so I adjusted his saddle and loosened the flank cinch which had caused him to buck in the first place. Two of the rail birds had come down off their perch and were sticking around close, waiting a chance no doubt to slip a cinder or a pebble under the saddle just to encourage the bronco to give a repeat performance. I bridled him and lost no time mounting. As nothing happened I spurred him and rode out of the yard at a smart trip, blowing kisses to the boys.

The Rainbow Grizzly

The crew I had that year were all green men except for Blackie, whom I had worked with in my first year with the Service. Moving day was always pretty much of headache until the boys learned to throw the diamond hitch. Our first job was to build a telephone line to Moberly Creek Lookout. This was a tree line and known as a ground circuit as only one wire was needed. Unfortunately the first mile had to have poles, as the timber was too small to climb. The wire had to be strung fairly high on account of the moose, for if they caught it on their horns they tore through the bush until the wire snapped.

We had to cross a muskeg also, which called for tripods, as it was too soft to hold poles. The boys were all good workers and the project did not take too long.

Next we moved to a camp about halfway between Winter Creek and Wildhay River Cabin where some bridges had to be repaired. One lovely morning we were going out to work when one of the boys noticed what he thought was a sorrel horse. He drew my attention to it and I saw right away that it was a very old grizzly. He was so old that his hair was a dirty yellow and, as we found later, several teeth were missing.

We sat down to watch him, as he was still some distance away. A creek meandered through the valley below us and the bear waded in and crossed to our side. When he climbed out he stopped and shook himself like a wet dog. The sun

Grizzly Bear.

FOOTHILLS RESEARCH INSTITUTE COLLECTION

Tripods carrying a phone line across a wet area. Bow River Forest, Rocky Mountains Forest Reserve, 1920s.

DOMINION FORESTRY BRANCH, ALBERTA FOREST HISTORY PHOTOGRAPHIC COLLECTION

The early Forestry bridges, such as the ones across the Wildhay and Berland Rivers, were typically built of local timber with log abutments and rock-filled cribs to support longer spans. The bridge in this photo is on the Highwood-Pekisko Creek road in the Bow River Forest, 1923.

DOMINION FORESTRY BRANCH, ALBERTA FOREST HISTORY PHOTOGRAPHIC COLLECTION

shining through the spray formed a miniature rainbow about him. It was a sight I will never forget. We referred to him afterwards as the Rainbow Bear.

He started coming straight towards us. I was afraid some of the boys might get jittery and run, as it was likely the first bear they had ever seen. When he was about a hundred yards from us, he stopped and listened. Some of the crew had started out ahead of us and he evidently heard them talking, for he suddenly wheeled round and started back in the direction from which he had come. I might say he didn't stop to shake himself when he recrossed the creek.

That evening some of the boys who had been out hiking or fishing said they had seen a campfire on the shore of a lake some distance away. I said it was likely some Indian out after beaver and gave it no more thought. A few days later I rode into town and the first man I met asked me if I had heard about Louie Shuswap the Indian being nearly killed by a grizzly. He proceeded to tell me the story. It seems that Louie had gone out after spring beaver and he had camped by the lake where the boys had seen the campfire. The following morning he had run into this grizzly and shot at him twice. The bear dropped and was to all appearances dead, but he was playing possum and the minute Louie got close enough to him he sprang up and commenced to maul and claw him. The beating and the shock put Louie out and he knew no more until his friends came looking for him. They found him and the bear lying side by side, the bear dead, and Louie almost so. After he had recovered, they asked him why he had not shot again and made sure the bear was dead as almost every hunter does. "No more bullet stop my gun," he said.

Wild animals have wonderful vitality and that bear must have mauled Louie even after he was mortally wounded. I knew of another case where a grizzly chased a hunter after it had been shot through the heart, and it looked like curtains for the hunter. Luckily, he kept his cool and shot the bear in the mouth, blowing most of the skull away. I had a look at the hide of Louie's bear and I was certain it was our Rainbow Bear. It is seldom that you see a bear whose hair is so bleached or faded. He must have been a very old animal.

Building the Trail to Donald Flats

After we finished repairing the bridges and putting in some much-needed culverts, we started on our new trail work. This trail was being made to open up the country lying between the Wildhay and Berland rivers. As no suitable ford on the Wildhay was to be found at the commencement of the trail, we were forced to travel downriver about two miles. A bridge had to be built across the same creek that our bear had crossed farther up. As it was not only deep but marshy on both banks, this called for quite a massive bridge with a long span and a truss. A pile driver would have simplified matters, but that was out of the question. We were camped there for quite a time but everybody enjoyed it. We had some wonderful fishing there and it was my first experience with the large-mouthed grayling. Rainbow and Dolly Vardens were plentiful but the grayling was the craftiest of the lot. We had a competition going and our cook caught the largest rainbow, Blackie the largest Dolly Varden and "yours truly" the largest grayling.

When we decked the bridge, a school of suckers seeking shade from the hot sun stayed under it. A sucker has a very small tender mouth, and after several futile attempts to hook them the boys rigged up a snaring device. They cut some long slender poles, then formed a noose of light brass wire and fastened it on the end of the poles. This was done only for sport as suckers are very bony fish and are seldom eaten.

When the bridge was completed, the trail cutting went ahead without incident. I had to make the trip to Entrance at the end of every month with the progress report, the time sheets and my diary and to bring back the mail and what supplies we needed. The boys were catching on to the packing, but I still did not care to trust them with a bunch of horses alone. The wily cayuse seems to know when a tenderfoot is in charge and there was always the chance of them missing the trail and getting lost. We had just completed the trail to Donald Flats, which was our objective, when I received instructions to cancel all future projects and return with the crew to headquarters. It was a bit of disappointment to all concerned, especially to the boys as the season was pretty well advanced and most of the summer jobs were taken and the winter bush work not yet due to start. But the powers that be had decreed it that way, and nothing could be done about it.

Dominion Forest boundary survey crew at their cook tent. Camp 1 along Athabasca River, Rocky Mountains Forest Reserve, 1911. This crew surveyed the north and east boundary from the Athabasca River to the 15th baseline.

DOMINION FORESTRY BRANCH, ALBERTA FOREST HISTORY PHOTOGRAPHIC COLLECTION

Hunting and Reconnoitring to the Smoky River and Sheep Creek

Cancellation of trail building sort of left me at loose ends also, and as I had been advised that extensive trail work had been planned for the next several years, I proposed to Tom Burrows that I would like to look over some of the country where we would be working next. Getting a rough idea of the route and blazing trails ahead would save tying up the crew at the time of construction. Tom thought it was a good idea too, so when Ernie Harrison was ready to start on his next patrol on the Mountain Trail, I accompanied him. A brother of Mrs. Burrows, Kenneth Campbell, came along for the trip too. Before I left I saw Bill Douglas and was pleased to hear that he was now an assistant ranger.[21]

[21] Ken Campbell was a forest ranger in 1926 and 1927.

The project that I wanted to look over was a trail from the Big Smoky River to Sheep Creek, thence to the boundary between Alberta and British Columbia. Ernie knew this part of the country fairly well, having at one time trapped there.

The weather was lovely when we started out and being with jolly companions, I thoroughly enjoyed the trip. We left the Mountain Trail at Big Grave Flats and headed for the Big Smoky River. The hunting parties and trappers occasionally went this way and they were clamouring for a decent trail and I did not blame them. It was simply terrific, especially after we reached Kvass Summit. From here

Dominion Forestry Branch pack string crossing the Kvass Summit, c. 1920s.

JACK GLEN PHOTOS, ALBERTA FOREST HISTORY PHOTOGRAPHIC COLLECTION

on to the Smoky, a distance of seven miles, you dropped over two thousand feet. We could have camped on the summit as there was horse feed of a sort there, but we decided to travel on to the river.

The scenery compensated for the poor trail – it was simply breath-taking. A lively little creek had its source at the summit and followed down a deep gorge, finally emptying into the Smoky. We followed this creek down, as it was a case of "Hobson's choice" with the mountains rising steeply on both sides. We made something like thirty crossings of that creek and were certainly relieved when we finally sighted the river. Looking at the lay of the land from our camp on the Smoky River and at the route we would take the next day, it certainly did not look very promising for a permanent trail.

Ernie Harrison mentioned that he would cross the Muddywater River the next day. As we were now close to the confluence of the two rivers, I asked him why the trail did not follow the Muddywater up to the point where it disappeared in a cleft in the mountains, as it was all grass covered with open side hills. He said he had never gone that way. I thought there must be a catch in it somewhere, and there was – but more about that later.

Stan Clark's Boat

The late Stanley Clark, a great friend of mine who outfitted and guided big game hunting parties, had packed a boat in three sections out to the Smoky River. That boat had had quite a journey before it reached the river, as the man in charge of the party had taken the wrong trail, doubling the mileage covered and having to cut a new trail part of the way. But, despite all the abuse, the boat was still quite seaworthy and we used it to cross the river the next morning as we did not want to get either ourselves or our packs wet. We swam the horses. The river was deep at this point and had been the scene of a drowning incident some time earlier.

We made a short trip the following day as we had to cut a lot of fallen timber out of the trail and Ernie said the next best camp was at Sheep Creek. This was known as the Goat Camp and was aptly named as we had been climbing steadily since we left the Smoky and were partway up a mountainside. The horses had to be content to graze on slough grass and the water was not so good either. We had what might be termed a late lunch and, having a few hours of daylight left, Kenny and I decided to take a goat hunt. Ernie said that he would stay and watch the camp, as we were right in the heart of grizzly country.

Soon after we reached the timberline, we spotted two fair-sized billies and started to stalk them. When we got within shooting distance, we decided on the goat that each would shoot at. Kenny had a Savage .22 high-power rifle, which I figured was a trifle light for big game, although I have heard of moose being killed with it. I had profited by my mistake the first season in not having a suitable rifle and was now armed with a .30 calibre U.S. army rifle, which I felt sure would take care of anything up to and including grizzly.

Stan Clark, George Bothwell and the University of Toronto Forestry Connection

The first forestry degree program in Canada was opened in 1907 at the University of Toronto. The Dominion Forestry Branch recruited many of its early graduates to staff its expanding system of forest reserves in the West. Foresters Stan Clark and George Bothwell were classmates in the graduating class of 1913, of which DFB hired nine of the eleven graduates. Clark was appointed supervisor for the Athabasca Forest at Entrance. Bothwell was first stationed in Ottawa but transferred to the Athabasca Forest in the spring of 1914 to work with Clark. They both later enlisted to serve with Canadian troops during the First World War.

In the meantime, DFB recruited Willis N. Millar in 1912 to be supervisor of all Alberta forest reserves. Millar was a 1908 Yale graduate who was forest supervisor with the U.S. Forest Service in Idaho. In Alberta, Millar focused on forest protection, especially developing a network of trails and cabins and construction of telephone lines. Among his notable achievements was a comprehensive survey and report on wildlife in the forest reserve that included a map of proposed game reserves, the locations of which influenced extension of the boundaries of Jasper and Rocky Mountains (Banff) parks. During his inspections in Alberta, Millar travelled with horses through the Athabasca Forest several times before he left in 1914 to teach at the University of Toronto. He shortly after served in the U.S. armed forces before returning to the University of Toronto where he remained on its academic staff until 1933. He would have known Clark and Bothwell.

Stan Clark returned to the DFB at Entrance in 1918 after the war, but at the end of July 1919 he resigned, homesteaded on what became the Athabasca Ranch and went into business for himself. He later sold the ranch to Harry Davison. Clark purchased the general store in Entrance from Bert Davey in 1946 and sold it to Gordon Watt in 1951. Clark's Crossing on the Smoky River was named after him. Unfortunately, George Bothwell was killed during the Battle of the

W.N. Millar, Forest Reserve Inspector for Alberta, on the Athabasca Forest 1913.

DOMINION FORESTRY BRANCH, ALBERTA FOREST HISTORY PHOTOGRAPHIC COLLECTION

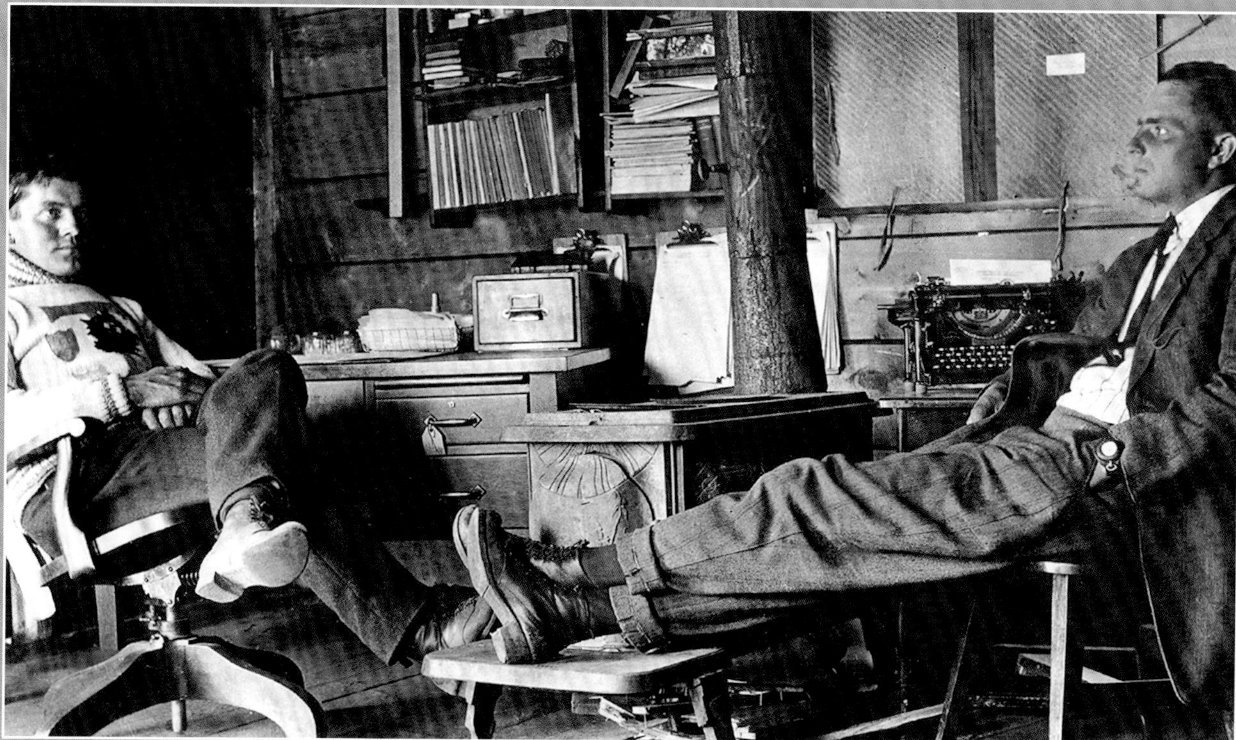

Stan Clark (l), Forest Supervisor Athabasca Forest with Charles Morse, Alberta District Superintendent. Hinton Ranger Station, 1913. Note the U of T Forestry sweater worn by Clark.

DOMINION FORESTRY BRANCH, ALBERTA FOREST HISTORY PHOTOGRAPHIC COLLECTION

Somme in 1916. The University of Toronto established a gold medal in his name for excellence in silviculture, his particular area of interest.

Also, in an interesting postscript, the eastern portion of Athabasca Forest in 1955 became part of the North Western Pulp & Power Ltd. lease negotiated to support Alberta's first pulp mill at Hinton. Des Crossley was recruited to serve as first chief forester. Crossley was also a University of Toronto graduate (1935) who had a special interest in silviculture and, after service in the Second World War, became a silviculture research forester with the Canadian Forest Service. He successfully pioneered the first large-scale commercial forestry operation that ensured forest regeneration and sustained-yield forest management at Hinton. Bothwell would surely have been pleased with the result. Crossley, in turn, hired Jack Wright (U of T 1953) who later succeeded him as Chief Forester, Bob Carman (U of T 1954), Russ Powell (U of T 1965), and Bob Udell (U of T 1966) and Paul Atfield (U of T 1974).

Sources: Hart, 1980; Graduate Directory 1992, University of Toronto Forestry Alumni Association; Historic Timeline, Foothills Model Forest History Project.

Kenny was to take the closest one to us as this was his first goat and I wanted him to be sure and get it. We both shot at the same time. I was almost certain that I had hit mine, but he disappeared behind a rock bluff and I lost no time getting to the spot where I had last seen him, as I hate to have an animal suffer. He was still alive, but another shot finished him off. As we needed some fresh meat for our trip I bled him, skinned out the cape and head and took as much of the meat as I figured I could pack down off the mountains.

This all took time and I wondered how Kenny had made out. He had done a lot of shooting and I wondered if he had had buck fever, that malady that affects beginners and causes their aim to be erratic. The shooting had stopped but I could see nothing of him until I heard him shout. He was a considerable distance away and waving frantically. I made my way down to where he was as fast as I could. His goat was still alive but mostly gutshot and all his ammunition was expended.

"For God's sake, shoot him for me," he said.

The goat was a long way off and the daylight was fading, which did not make conditions too good. First he would stagger along for a bit, then he would fall down and roll down the mountain until he fetched up against some obstruction, and do the same thing all over again. It took me three shots to finish him off and then I told Kenny to skin out the cape and to be sure and slit along the top of his neck to the shoulders, otherwise the head would be spoiled for mounting. I showed him how I had done mine and left him to it as I wanted to get back to

Mountain Goats in high country.
ISTOCKPHOTO

View of Dry Canyon from the east side, as it slopes north towards Sheep Creek, c. 2005.

SUSAN FEDDEMA-LEONARD AND THE WILLMORE WILDERNESS FOUNDATION

camp with my load before it got too dark. Kenny would have no load to speak of and would make better time.

I got down about fifteen minutes ahead of him. Ernie had a pot of cocoa simmering by the campfire – the end of a perfect day. Those two billies must have been twins as they both measured exactly the same and looked lovely when mounted.

The travelling was exceptionally bad the next day. As I mentioned, we had camped at the edge of a slough and had to follow it along for a mile or so with the ground getting softer all the time. I was afraid some of our horses would go down, but luckily we at last got on to better footing. We went down a long, very steep slope into the Muddywater valley. After crossing the river we climbed again until we reached what is known as the Dry Canyon. The going was good here except that there were lots of windfalls, and at last we cut only the worst of them as we were losing too much time. The canyon ended at Sheep Creek and there we camped.

The weather had been so nice when we started out we hadn't bothered with a stove. We had a small tent but had never used it so far, so after supper I caught a nice bunch of Rocky Mountain whitefish, picked a friendly spruce, unrolled my sleeping bag and went to bed. Sometime towards early morning I woke and

Rough Camping in early snowfall, c. 1940s.

JACK GLEN PHOTOS, ALBERTA FOREST HISTORY PHOTOGRAPHIC COLLECTION

looked out. Imagine my horror to find the ground covered with snow, and it was still falling heavily. There was no use worrying about it, so I pulled a canvas cover over my head and went back to sleep.

What a mess there was with everything covered with snow when we finally crawled out in the morning. We had no thought of moving that day, so we pitched the tent. That was when we really missed a stove as it can mean all the difference between comfort and discomfort.

After lunch Ernie and I took our rifles and hiked up the valley. We climbed a likely looking mountain and were rewarded by seeing a large number of goats.

The storm had driven them down to the lower levels and, no matter which direction we looked, we could see goats and more goats. In this part of the country there are a number of black shale slopes that the goat often frequent. This makes a mess of their white hair, but it also made them more easily seen than against the snow. We also saw a number of caribou. They have a peculiar way of walking – a lazy sort of slouch – but when they run they are very graceful and the action resembles a thoroughbred hackney's gait. Ernie thought he would try for a goat, but all I was interested in was grizzlies. He shot one but took only the hide and a little chunk of the meat. This is practised a lot by big game hunters who are after bear. They will kill an animal and only take the head and sometimes the hide, figuring on returning a few days later to perhaps surprise Bruin feasting on the carcass.

The snow quit and the weather turned very cold, so much so that I had to chop through the ice on the creek the following morning to get water for breakfast. After chopping a supply of firewood we crawled into our sleeping bags, as that was the only way to keep warm. We had brought nothing to read, and besides it was too cold to read.

A Welcome Encounter

By the third day, this routine was becoming rather monotonous when we heard the sound of an axe some distance downstream. This meant we had neighbours so we went visiting. We found the camp about a mile away with a party of three, an elderly hunter from the United States, a horse wrangler and a combined guide and cook. They made us very welcome and we were invited to stay for supper. It made a pleasant break for us, as it was very comfortable in the tent with a roaring fire in the stove. They had bagged a moose so far and the hunter said that if they got a fair-sized ram he would be satisfied.

Bull Moose.
ISTOCKPHOTO

Shortly after supper our host excused himself, saying that he was a bit weary and was getting too old to climb mountains. After he had gone to his tent the guide told us he was rather eccentric about his food and after each meal washed and scalded his own dishes. His plate was a double-decker affair, the idea of this being that the food could be kept hot by filling the lower compartment with hot water. Now Charlie (the guide) took quite a fancy to this plate and ate his breakfast off it, unknown to the old man who did not get up very early. "Of course I washed it," he said, "but it's not always scalded and I don't use a fresh tea towel every time, either."[22]

[22] This may well have been Charlie Matheson, park warden and guide. His widow Mona (Harragin) Matheson said the story sounded a lot like Charlie.

Their camp was near Famm Creek and as I got better acquainted with the country I noticed that this was the limit of the sheep range. The country beyond was more rugged and more suitable for goat and caribou.

Next morning we took our rifles and rode our horses up through the pass, as we could see no point in camping beyond this late in the season. On the return trip we tied up our horses and climbed up to where Ernie had shot the goat, but

Rangers with Pack String on Switchback. Brazeau Forest 1915.

DOMINION FORESTRY BRANCH, ALBERTA FOREST HISTORY PHOTOGRAPHIC COLLECTION

nothing was left but bloodstains in the snow. The wolves had beaten the bears to it. That evening a Chinook wind sprang up and the next morning the snow began to disappear so we headed back to the Smoky.

When we came to the ford on the Muddywater River, I decided to follow downstream and see if a detour of the swamp was possible. I tied up the bridle reins on my saddle horse and let him go down with the others. It was arranged that I would contact the others again on the trail at a point between the Goat Camp and the Smoky River. I soon realized that a trail there was out of the question as the river flowed through a deep canyon. In fact I had a hard scramble getting back to the trail, by which time it appeared like a boulevard to me.

We spent a day at the Smoky River. The weather was nice again so we rode down the river for a few miles, picked some blueberries and returned. I now had a pretty good idea what I was going to be up against cutting trails in that part of the country. I made up my mind that the sooner another boat could be brought out the better because then it would be possible to have a boat on each side of the river. This we eventually had. We had the luck to find the boat on the right side on both occasions.

We stayed two days at Rock Lake to rest the horses and ourselves since we had covered over two hundred miles of rough mountain trails. On the way home to Entrance we took a different trail that shortened the distance by at least two

Right: Brewster's Original Solomon Creek Cabin remains in use by Skyline Trail Rides which has added other buildings on the site, c. 2000s.

TOM VINSON JR.

Old coal mining exploration building, Thoreau Creek, 2003.

BOB UDELL

Old Steam Engine (1928). Located just north of old mine buildings and used for coal exploration. This mine prospect was never developed. Thoreau Creek, 2003.

BOB UDELL

miles. A coal mining company had cut a sleigh road to what was known as Carson Creek and hauled in a diamond drill to prospect for coal. It was a very good trail and later on Major Fred Brewster from Jasper built a cabin on Solomon Creek to be used by his trail riders and horse wranglers. The door was never locked and in later years I travelled that trail and stayed in the cabin if it wasn't occupied. There was excellent fishing in Solomon Creek. Montana Pete, too, had a cabin on the summit between the headwaters of the south fork of the Wildhay River and Solomon Creek.

Fred and Jack Brewster

Fred Brewster (1884-1969) operated a freighting team and packing business out of Prairie Creek in 1910. In 1912, he and his brother Jack moved to Jasper, setting up as packers and outfitters; Jack was 17 at the time. Fred Brewster later built guest lodges at Maligne Lake, Solomon Creek (now the Black Cat Ranch) and in the Tonquin Valley. As well, he built the Skyline Trail, a project that took over four years. In 1925 Fred organized the Trail Riders of the Canadian Rockies, based in Jasper. Brewster's Wall in Willmore Wilderness Park is named after Fred and Jack's camp there. Jack Brewster opened his own business in 1922 and became renowned for his guiding skills, especially on big game hunts.

(Hart 1980, Feddema-Leonard 2007, Karamitsanis 1991)

Winter at Entrance Headquarters, 1923–24

The fire season was almost over so Ernie Harrison and I took the horses that would no longer be required back to Entrance. We also gathered the horses from the Brazeau Forest Reserve that had been brought over to be wintered. There was no hay for them there unless it was bought, and this would be an expensive business.

The Brazeau people had a mule in their pack outfit and the horses bullied him. Moke was his name, so when I fed oats I would call him by name and put out his pile a little distance from the rest. Mules are slow, dainty eaters, so when one of the horses had finished his own portion and spied Moke still eating, he would put his ears back and charge towards him. But Moke would be watching out of the tail of his eye, and his jaws would be working overtime. He would wait until the horse was within kicking range and then let him have both barrels. I never knew him to miss.

Some of Moke's actions were almost uncanny. For several years he had led the Brazeau horses back to Entrance in the spring, almost to the day, and not until the feed was well started. The Brazeau rangers claimed that he could walk a log across a ravine and I didn't doubt it. One time we had the Brazeau horses rounded up and were waiting for the rangers to show up so we could help them to get their stock started in their swim across the Athabasca River. Moke was doing a little investigating on his own and found his way to the garbage dump. He managed to get one of his forefeet stuck in an empty can and nearly went crazy. It took all hands to get that can off as he could not be caught and could dodge a loop of a lariat like nobody's business. After he was roped, that was only half the battle. He bit and kicked at anyone who approached him and it looked as though chloroform was the only answer, but like the monkey and the chestnuts, who was going to administer the knockout drops? We did not want to hurt the little fellow, so finally we succeeded in getting a second rope on him and throwing him. Then by sheer force of numbers we subdued him and removed the can. He was still fighting mad when we let him up, as I suppose he had never been so humiliated in his life.

Moke, the trusty pack mule, Brazeau Forest, Rocky Mountains Forest Reserve, 1913.

DOMINION FORESTRY BRANCH, ALBERTA FOREST HISTORY PHOTOGRAPHIC COLLECTION

Fording the Athabasca River at Entrance, c. 1920s. The Canadian Northern Railway ferry at Prairie Creek may have been installed as early as 1911.

JACK GLEN PHOTOS, ALBERTA FOREST HISTORY PHOTOGRAPHIC COLLECTION

The winters at Entrance were far from being as dreary as a person would suppose in a small place like that. We had daily train service, good skating when the river froze over, the pool table and dances galore. When the Indians came in to sell their furs and trade, they presented quite a colourful spectacle, with the men in their buckskin clothes and the women in all their brilliant finery. At that time, lemon extract was much more potent than it is now and they bought it in large quantities. Bay rum, painkiller and Canned Heat were also in great demand. They had their brawls but they never bothered the white population.

Radio at that time was in its infancy and I had a four-peanut-tube set. Night reception was all that could be had and we used headphones. Later, when screen grid tubes came into being, we could get reception in daylight, but the speakers were separate. I traded in the old one and got one of these sets. The Indians didn't take too kindly to the radio sets at first, being a bit afraid of them, but they liked phonographs. One old fellow who figured he was a whole lot smarter than the rest of the tribe told them they had nothing to be afraid of as the Monias (white man) was just fooling them and had a bunch of little records inside the box.

Johnnie, this same smart Indian, brought a friend into the cabin once during the winter to see me about a horse deal. My radio speaker was a large horn very much like the one in the "His Master's Voice" advertisement and I had an extension cord from the set to where it sat on a small card table. Johnnie's friend was sitting with his back to this table and without thinking I walked over to the set and switched on the volume. Being battery operated the sound came through instantly and a man with a deep voice began to make a speech. If ever there was a badly scared fellow it was that one. He let out a yell and bolted for the door and no amount of persuasion could induce him to come back.

A Popular Priest and the Home Remedy

The Indians in that district were Catholics and periodically their priest visited them. His name was Father Beaudry and he was liked by white men and Indians alike. In the evenings Ernie Harrison and I often played pool to while away the time. One night Father Beaudry was staying at the rooming house, so we coaxed him to try his skill at pool. He had us in stitches most of the time with his antics. From then on, we made a point of seeing that he won by switching the high numbered balls in his favour. After a game or two he needed no inducement to play and we looked forward to his visits with pleasure. After awhile he began to smell a rat with all his phenomenal luck, and in accented English remarked that there was something rotten in Denmark. The way he said it kept us laughing for long enough.

One time, Father Beaudry did not feel like playing pool as he had a terrific cold. The stock of drugs at the little country store was somewhat limited. Everybody sympathized with the Father but it was Ernie Harrison who came up with what turned out to be the cure. He told the priest to get into bed, saying that he would either cure him or kill him. "It doesn't matter," the priest said, "I feel as though I'm lying on the brink of Hell now." Ernie in his most professional manner handed him a vile concoction consisting of oil of peppermint, sugar and very hot water. The tears ran down the old man's cheeks but he drank it to the last drop. When we visited him next morning, he was quite cheerful. I asked him how he liked Dr. Harrison's prescription. He laughed and shook his head. "The Devil himself was lurking in the brew," he said, "but between him and Harrison, they cured me." I never did find out what became of Father Beaudry after he was transferred, but his successor was a different type of man entirely.

A Visit with Montana Pete

A lot of our time in winter was occupied in woodcutting as everybody on the Forestry staff had wood-burning stoves. We wanted to clear a piece of land in order to grow green feed for the work teams. This land had white poplar growing on it so when it was split and dried it made wonderful cookstove wood. For the heaters we used spruce. When we went out to cut spruce, Ernie and I would pack a lunch kit, as it was too far to come home. One day a cold wind was blowing, so Ernie suggested that we go down to Montana Pete's place and eat our lunch there. We figured he must be home as smoke was coming out of the stovepipe.

Entrance Headquarters and bunkhouse cabin showing woodpiles, 1930s

JACK GLEN PHOTOS, ALBERTA FOREST HISTORY PHOTOGRAPHIC COLLECTION

Pete was quite a character all right. His shack was a mess, with saddles and saddle blankets lying everywhere. The table was littered with dirty dishes, empty cans and stale bread, but the cabin was comfortable even if it did smell somewhat. Unable to find a clean spot anywhere I nursed my lunch kit on my knee while I ate. Ernie did the same. If Pete noticed this he did not say anything.

He was too busy storytelling – and what stories he did tell. He had travelled the length and breadth of Western Canada and the United States and had seen a great deal of lawlessness and life in the raw. His exploits were so interesting that we were held spellbound and the first thing we knew it was time to go home. He invited us to come back, but we never did as we figured we wouldn't get much wood cut that way.

Pete kept the wolf from his door by trapping in the winter and guiding the occasional summer tourist party. He had tremendous stamina. Once he froze one foot while on a trapline eighteen miles from Entrance and no doctor closer than Jasper some fifty-odd miles away. When he realized that he was going to lose some of his toes and probably his foot he snowshoed into Entrance in below-zero weather. The torture that he must have endured would have driven most men crazy. He died a few years later of pneumonia in the hospital at Jasper, cheerful to the last.

Harry and Curly Phillips

I took over the teamster's job in winter and this entailed quite a bit of work. There was wood to be hauled as well as building logs and fence posts. Baled hay had to be brought in from Winter Creek. The winter road followed along swamps and crossed a lake known as Jarvis.

The hay haul was commenced as soon as the ice on the lake was considered safe for travel. Ernie went with me on the first trip, and, crossing the lake, walked ahead, keeping a sharp lookout for air holes. I would leave Entrance in the morning and forego lunch until I arrived at Winter Creek, generally about 2:30. After lunch I loaded up for an early start the next morning. Except for the cold crossing the lake, I enjoyed the work and kept at it until I filled the hay shed with about twenty tons.

It was on this job that I first met Harry Phillips, who had a commercial fishing licence and used to supply Entrance and Jasper with whitefish. He and his brother Donald (Curly) outfitted and guided hunting parties in the fall and sightseeing parties in the summer. They preferred canoe trips to horse trips, but they were experienced in both. I never got to know Curly too well but heard about some of his exploits.

On one occasion he had a party out in the mountains in the charge of a Jasper guide. A message had to be delivered to one of the party and as no one was available in Jasper, Curly started out on foot to deliver it himself. In due course he reached the hunting camp and stayed overnight. The following morning, the

Curly Phillips (1884-1938)

"Everyone held the same high opinion of this man, who in all things was simple and straightforward." (Curly's biographer, Dr. William C. Taylor) Donald (Curly) Phillips was a legendary Jasper guide and outfitter who, as biographer William C. Taylor described him, was "one of the colourful band, now sadly diminished, who worked hard to open up the mountain trails in the first half of this century." He came west from Ontario in 1908 and settled in Jasper. Among his most heroic achievements was the first, almost successful, attempt with Rev. George Kinney to climb Mount Robson in 1909. He became a preferred outfitter for many expeditions in the Jasper area as well as the old Athabasca Forest Reserve, including famous clients such as Baron Byng of Vimy, Governor General of Canada from 1921 to 1926, and Kenyon V. Painter, an American millionaire whose photos of a Phillips-led 1916 hunting trip on the Smoky River are preserved in a the Bruce Peel collection at the University of Alberta. Phillips Mountain, across the Robson River from Mount Robson is named in his honour.

Sources: William C. Taylor, *Tracks Across My Trail: Donald "Curly" Phillips, Guide and Outfitter,* Jasper Yellowhead Museum and Archives, 1984; and Dave Birrell's Peakfinder website, www.peakfinder.com.

Curly Phillips (l) with Charlie and Mona (Harragin) Matheson at Circle "M" ranch, early 1940s. Mona (Harragin) and her sister Agnes were the first licensed female guides in Jasper National Park. Agnes married Mark Truxler.

MARK TRUXLER COLLECTION, COURTESY OF JACKIE HANINGTON

horse wrangler asked him which horse he wanted saddled for the return trip. "Oh I don't want a horse," he said, "I'm in a hurry." A favourite saying of his was, "If you want to be sure of getting somewhere – walk."

On another occasion he had been treed by a grizzly and was becoming rather impatient as the bear showed no signs of leaving. He was just getting ready to come down and do battle with his hunting knife when the bear decided to move off. I suppose Curly was in a hurry that time also.

He was absolutely tireless himself and was hard on his horses as a result. In every pack train – and Curly's was no exception – you would find one or two foxy horses who would battle their way into the centre of the pack train and hold up the works. On narrow trails nothing could be done about it as it was impossible to get at the culprits. Curly hit on a solution for this. He bought an air gun and peppered the miscreants at will. This method backfired after a while, though, for when camp was reached and somebody stepped on a twig the whole outfit was so jittery it was almost impossible to catch and unpack them.

Several years ago Curly and a teenage boy were swept to their deaths by a snow slide while skiing on Pyramid Mountain near the town of Jasper. It seemed like both of the brothers were destined to die with their boots on – but more about Harry later.

Building the Athabasca Lookout, 1924

It had been decided to build a lookout cabin about six miles from Entrance. We decided against building a tower as that would have meant also building a cabin for the lookout man. This was a problem as it was a very stiff climb and helicopters were as yet unknown. A few lookouts had already been built in Alberta, but at a terrific cost. Tom Burrows figured that if we could build this one fairly cheaply, the money saved could be spent on the building of much-needed trails. The usual procedure employed in the building of the other lookouts had been to pack the lumber on two horses in tandem, but this was hard on the horses, extremely dangerous and expensive.

Several suggestions were put forward including cutting the lumber in short enough lengths for one horse to pack. I ruled this out; a building of that sort would have to be strongly built to withstand gale-force winds and it would break a carpenter's heart to build it with short boards. Another scheme was to skid the lumber the way the Plains Indians moved their belongings – on a travois. This too was rejected, as the ends of the boards would be split and worn off in their passage over the rocks. The south route was far too steep to use sleighs and besides it was bare much of the way.

I hit upon a scheme that I thought might work. We would use the north slope of the big ridge and by following a ravine that led in the right direction, I would be able with very little work to get a passable sleigh road up to the foot of the hill. I made several trips with the same team and sleighs and got all the

A gruelling duty - horses were used to drag lumber for construction of the Coliseum Lookout near Nordegg, Clearwater Forest, 1927. Moving construction materials before the advent of dozers and helicopters required ingenuity, strong backs and considerable horsepower. Athabasca Lookout was built with materials hauled up in a similar way.

DOMINION FORESTRY BRANCH, ALBERTA FOREST HISTORY PHOTOGRAPHIC COLLECTION

material relayed to that point. Then Tom Burrows, who said he wanted to get away from the office for a few days, came up with me and we made camp and erected a shelter for the team. He was sceptical about my route and could hardly be blamed, as the north slope looked very formidable.

This was all taking place in March of 1924 with frosty nights and bright sunny days. We had to shovel snow when we pitched our tent, but with a good covering of spruce boughs on the floor and spruce boughs to sleep on, we were quite comfortable. The cook stove furnished the heat and the Coleman lantern helped also. At nights we put blankets on the team so they did not suffer either.

After we had made the camp shipshape I explained my plan. We would use only the front bobs of the sleighs and pile the lumber with the rear end dragging in the snow. As the north slope was shaded we would have snow all the way to the top for at least another month. The first part of the climb would be the worst, although not the steepest, as a number of large windfalls had to be cut out. We also tramped the snow down as a harnessed team in deep snow is inclined to flounder. We were a tired pair that night but decided to start hauling the next morning.

We could not expect the team to pull much of a load until we had a trail broken, so I piled on just enough lumber to cover the bunk and we started out by easy stages of about a hundred feet followed by a short rest. We kept this up till we reached the spot where we quit the night before. Then the real work began,

Athabasca Lookout, 1942. Fred Hendrickson was the lookout man.

ALBERTA FOREST HISTORY PHOTOGRAPHIC COLLECTION

as we now had to cut trail and tramp snow ahead of the team. Tom must have smoked a lot of tobacco that day for every time I went back for the team he rolled another cigarette.

We avoided side hills as much as possible, so we were forced to go straight up some mighty steep slopes. When I stopped the team, the back of the load would dig into the snow and then the horses could rest without having to hold the load from slipping back. When noon came, we were both tired and hungry, but as we were almost two-thirds of the way, we decided to go on to the top. We finally made it, but after we unloaded we spent considerable time digging up enough rocks to weigh down the lumber, as the winds at that high altitude are terrific. The team by this time were cold and as soon as they had their heads towards home, I had to hold them in for all I was worth to keep them from galloping, as our seat on that narrow bunk was none too secure. To celebrate we declared the next day a holiday.

Tom went with me on the second trip just in case anything went wrong, but from then on he stayed in camp, cut wood and cooked our meals. Every trip, as the trail improved, I increased the load and took a few bundles of cedar shingles on top. Towards the last, I could make the trip in a little better than four hours and came back in twenty minutes. Then at night in camp I read and Tom played solitaire, as I never cared much for cards.

After the lumber was all hauled, I went out to Winter Creek to relieve Bill

Early Lookouts

A prime duty of early Dominion Forestry Branch rangers was to prevent forest fires, keep a lookout for fires and suppress them when found. As Jack Glen's diary shows, they took this responsibility seriously. Small wonder, as the landscape they encountered in the foothills of Alberta was characterized by the bleak aftermath of catastrophic fires in the late 1800s that consumed almost forty percent of the standing timber blanketing the hillsides. Early rangers did not have the benefit of a series of lookout towers such as we have today, and often "made do" with improvised lookouts such as this one used by Ranger MacDonald of the Brazeau Forest, south of present-day Hinton, around 1915.

Ranger MacDonald, Dominion Forest Ranger, on an improvised lookout near Coalspur, c. 1912.

DOMINION FORESTRY BRANCH, ALBERTA FOREST HISTORY PHOTOGRAPHIC COLLECTION

Douglas, who had been feeding the pack horses during the winter. I hauled out a load of oats as the hay was running out and spring hadn't arrived yet. Two deer and two moose showed up for their daily ration of hay. The horses didn't mind sharing their hay, but when it came to oats the four visitors were chased away.

During my stay at Winter Creek I cut a pile of firewood and fence posts until it was time to pick up my crew and start our trail work. However, it was decided to postpone the cutting of the Sheep Creek and Smoky River trails for another year. Since we had the lumber and material for the Athabasca Lookout on the site, we might as well get it built and manned.

A Scottish carpenter, Angus McIvor, was hired to do the work, assisted by Ernie Harrison. Angus hated the assignment and maintained that there was a feeling of evil about the damned place, but this did not in any way deter him from doing a first class job. I myself have shared the opinion of others that the environment seems charged at high altitudes. A person will jump at the slightest sound. It does not, however, affect everybody alike.

Since we were not going to build a separate house for the lookout man, everything had to be arranged so that his view would be unobstructed. The table had to do double duty. He had to eat off it and have his maps on it, as well as his telephone, for two-way radios had not yet come into being. The table had to be in the centre of the building with the wires to the telephone coming up through the floor.

The lookout resembled a greenhouse with glass on all four sides. Guy lines from the eaves on all four corners were necessary to hold the building down. Holes were drilled in the rock for the guy wires and leaded in. All four wires had turnbuckles. In building the telephone line, we were fortunate to find trees to hang the wire on, except for a rocky hill where no timber grew. On this stretch we used tripods, as we could not dig holes for poles.

The Athabasca Lookout was finally finished and Tom Burrows then advertised for someone to man it. The first applicant was nicknamed Lord Plushbottom. He resembled that gentleman and was just about as brilliant.

A New Trail Season Begins, 1924

Our first main project of 1924 was the bridging of Orchard Creek. It had long been a troublesome spot as only light loads could be brought across it. Lumber for its construction was out of the question so we used logs for caps and stringers and poles for decking. The poles were adzed off wherever they appeared to be too high, to lessen the bumps in crossing. We needed a large fill of gravel at each approach and this took a long time to gather, as we had nothing but horse-drawn scrapers to do it with.

We did a lot of maintenance also along the trail to Moberly Creek. The summer of 1924 was unusually hot and from one of our camps going to work we crossed a lovely little creek where all hands stopped and drank and filled their water bags and jars. One morning we found that it had dried up and everybody was greatly disappointed. We had old Johnnie, the Indian, on the crew, and when he saw the dried up creek he shook his head. “Twenty years no seeum like this before,” he said sadly. A short time after that one of the fellows, Fred Hendrickson, asked Johnnie how long he had lived in this part of the country. Johnnie thought for a moment. “I guessum twelve years,” he said. Everybody had a hearty laugh and though I don’t suppose Johnnie saw the joke he laughed too, to be sociable.

Fred Hendrickson's Colourful Career

Fred Hendrickson was a big, good-natured fellow who had been a whitewater river driver in his younger days and could keep us spellbound with stories of his experiences and near drownings. Once he was travelling with their wanigan, which was a lumberman's boat for carrying tools and supplies – they also ate and slept on it – when he noticed a large sign nailed on a tree: "Meals, fifty cents . . . Square meals one dollar . . . Immortal gorge meals one dollar and a half."

They were all fed, he said, but had no great variety. The principal item on the bill of fare was salt pork, which Fred referred to as the "long clear." This was what we now call back bacon, but different from ours in that it was dry cured and was in long strips from overweight hogs. He still liked it and when he was batching he was never without it.

Fred Hendrickson and Family, c. 1940s.

FRED TIETGE COLLECTION, COURTESY OF STAN TIETGE

Fred Hendrickson (1869–1962)

Fred H. Hendrickson (1869-1962) was a long-time resident of Entrance. Born in Minnesota, he was a "barker" for the *Barnum and Bailey Circus* before finding his way to Alberta. He was listed as a DFB forest ranger on the Athabasca Forest in 1924. He seems to have preferred seasonal employment that enabled him to trap in the winter. He helped cut the trail from Entrance to the Smoky River and ran the checking station at Entrance in the fall. He was most renowned as a fire lookoutman – both for his keen vision and knowledge of the country. He was at the Grave Flats and Lovett lookouts but best remembered for his 30 years at Athabasca Lookout. His cabin and trapline was at Rock Lake. Hendrickson Creek was named after him. He died in Entrance at age 93. Hendrickson Creek along Highway 40 is named in his honour.

(Feddema-Jones 2007, Gilliat 1998, Hart 1980, Karamitsanis 1991)

Isaac Walton (1593–1683)

Isaac Walton was an English writer and fishing enthusiast, the author of *The Compleat Angler* (1653). After the printing of this famous treatise he retired from business and spent the last years of his life travelling here and there, visiting eminent clergymen and others who enjoyed fishing, compiling the biographies of congenial spirits, and gradually enlarging on his treatise in five subsequent revisions.

There is hardly a name in English literature whose immortality is more secure, or whose personality is the subject of a more devoted cult. *The Compleat Angler*, with over 300 editions issued, is one of the three most published books in English history, the other two being the Bible and the complete works of Shakespeare. The Isaac Walton League continues today as a major conservation organization.

Source: Andrew N. Herd, *Fly Fishing History 1993–2005*, available online at www.flyfishinghistory.com/walton.htm.

One night in camp we were discussing the chariot race in Ben-Hur. Fred, who had been reading a magazine suddenly spoke up. "I was with Ben-Hur," he said.

"You must be a much older man than I think you are, Fred," I said. It transpired that he had worked as a roustabout in a travelling circus when they played Ben-Hur.

I was fortunate to have Fred's services for a number of years on my trail crew and he got to know the country like the palm of his hand. In later years he applied for and got the job of lookoutman at the Athabasca Lookout. Whenever Fred reported a fire to me, he gave such a good description of the location that I never had the slightest difficulty finding it. It was said of Fred that if a man lit his pipe in the Athabasca Valley within his range of vision he would see the first puff of smoke. This might be a slight exaggeration, but there were few better than him.

Trout Fishing Derby on the Oldman Creek Trail

We cut a trail to Oldman Creek that season too. This was not only for fire protection, but also to make accessible the area adjacent to a valuable block of spruce timber. I had my first taste of wild huckleberry while blazing that trail and I got all the crew out that evening after supper. We picked berries until almost dark and then as long as they lasted we certainly had some toothsome pies.

Our cook, Don Empson,[23] was as good a trail cook as you could find anywhere. There was a saying at that time, rife with disgruntled trail parties, that all good trail cooks had been killed in the war and the ones who survived should have been. However, Don had come through the war and I never once heard any complaint from the boys about his cooking. When I first met Don I thought what a cranky looking character he was, but my first impression was wrong as he was one of the jolliest of men.

As I mentioned, this was a particularly hot summer, and as we left for work in the morning Don would be standing by the cook tent. He would look up at the sky and shake his head. "I'm afraid it's going to rain, boys," he would say. Then every man went back and grabbed his coat. This kept up for several days with no rain or any sign of it and one of the boys finally told the cook that as a weather forecaster he was a fizzle. Don laughed, "I didn't say *when* it would rain," he said, "and it always has rained some time or other, hasn't it?"

We camped on a creek that was literally full of rainbow trout and someone suggested that we should have a rainbow derby. The simple rules were agreed upon and the prize money deposited. A real Isaac Walton enthusiast challenged me, and a judge was appointed to start us off.

We flipped a coin to determine who should fish upstream and who downstream, and the stream was declared out of bounds to the non-contestants in case they might be tempted to slip a fish or two into their favourite's basket. We were

[23] Don Empson was a camp cook, one of the best as described by Glen. A World War I cook, he is also mentioned in People and Peaks as a camp cook on fall hunts in the 1920s. (Feddema-Leonard 2007)

to leave camp on the stroke of a certain hour and fish for an hour. The man with the greatest number of fish would win. I returned on the hour with ten trout. My opponent arrived fifteen minutes later with twelve, but he was disqualified. We might have had another contest but by then it was time to move camp.

At another camp we found a colony of beaver working, so we decided to try to get a photograph if we could catch them in action. The damming of the creek had formed a small lake around which all hands and the cook deployed and waited patiently for the beaver to put in an appearance. Everybody kept still, but time went on and there was no sign of our friends. My foot had gone to sleep and I cautiously eased up to change position. That did it. A beaver noticed the movement and slapped his tail on the water to give the alarm. We knew then there was no use in waiting any longer. We might have seen some action if we had broken their dam but nobody had the heart to do so and besides they were causing nobody any harm there.

Building the Adams Creek Trail

Our next move was over to the Big Berland country to begin work on a trail from our Lower Trail to Adams Creek. My friend Bill Douglas joined us here as he had his district in pretty fair shape and wanted to get away from his own cooking. He and I both had special hunting licences as well as the resident's big game hunting licence, so he suggested that we try for a moose to make a change. We welcomed this suggestion as a steady diet of bacon and canned meat can get awfully tiresome.

Bill hunted for a few days and finally shot a nice fat young bull and asked me if I would go along with him and help pack home the meat, as it was too difficult to use a pack horse. He had just brought back the heart, tongue and liver and we had liver and onions for supper that night, which was a real treat.

We decided to go after it that evening, as it was risky to leave meat overnight on account of wolves and bear. The weather was not looking too good and, although it was dry when we left camp, it was not long before we had a stiff hailstorm. We sheltered under a friendly spruce tree until it stopped. Then Bill had trouble finding the moose and we had to wait out another hailstorm. When we did finally locate the moose it was almost dark. We loaded all the meat we could carry on our pack boards and started back.

Bill had the reputation of being a first class man in the bush and I let him take the lead and paid very little attention to where we were going. It was now pitch dark and it began to rain. We had been following a game trail which, Bill said, was heading in the right direction and was just barely visible, and then according to him we should leave it pretty soon and head more to our right.

Then our troubles really commenced. We fell over logs, sank knee deep in bog holes and almost had our eyes knocked out with limbs and branches. I imagine we started out with about sixty pounds of meat each, but by this time it felt closer to eighty or ninety. The shoulder straps were beginning to chafe badly when we hit another game trail.

"Now we'll soon be at camp," said Bill, but I was beginning to lose faith in his leadership. It seemed to me we had been over this trail earlier and I called a halt. Bill laughed at my suspicions, saying I was imagining things because of my fatigue. I was not convinced so I walked ahead a few steps, got out my lighter and examined the trail. Sure enough there were our tracks. We had been walking in circles.

"This is where we camp till daylight," I said. Bill said we need to carry on because it was too wet to start a fire in the rain, yet without it we would freeze to death. "I'm going to attempt it anyway," I replied. There is a sort of dry moss that grows on trees that the Indians call "Jackpine whiskers," and I have been told the caribou exist on it during the winter. I had this in mind to start the blaze and with the light from it I could possibly find dry twigs to build it larger. After a struggle we got a fire started and then we were able to find some dry limbs to build it up. In no time it was sending out quite a heat. I gathered some moss from under the trees and although it was damp it was more comfortable than the bare ground. The rain stopped about midnight and the stars came out but it also started to freeze. I found an upturned stump, which I dragged over, and this kept the fire burning for the rest of the night. I don't know how Bill made out but I managed to get some sleep. But first I studied the sky and located the North Star so that I knew what direction we must head for camp.

As we started out the next morning I heard the faint tinkle of a horse bell and then we had no trouble getting back to camp. Bill felt pretty bad about getting me into that mess, but it taught me a lesson which I never forgot: always pay attention to where I was going.

A few days later while locating and blazing trail, I saw a she-grizzly and two year-old cubs. I certainly didn't want any argument with that trio so I sat down and waited until they got out of sight. She was a monstrous animal and the biggest live bear I have ever seen before or since. The cubs were quite big too, and after their winter hibernation would be on their own next season.

The Pack Rat Hunt

With the completion of trail building and maintenance for the season, it was again time to head back to Entrance. To save pitching two tents we decided to pull into Grande Cache Cabin and stay there for the night. As soon as I opened the cabin door I knew we had company. A pack rat jumped on to the windowsill and made his way out through a broken pane of glass. It was a bit dark by this time and Don, the cook, got a magazine out of one of the packs and blocked what he thought was the broken pane with it. We were lying on our beds after supper when we heard a noise at the window and in popped the rat. Don had covered an unbroken pane.

Equipping a trail crew for a summer's work took a lot of supplies and gear. This photo shows Jack Glen (to right of cabin door) and some of his crew with more than a tonne of supplies for trail work. Rock Lake Cabin, 1939.

JACK GLEN PHOTOS, ALBERTA FOREST HISTORY PHOTOGRAPHIC COLLECTION

MAGIC
BAKING
POWDER

Forestry pack string at Muskeg Cabin, c. 1930s.

JACK GLEN PHOTOS, ALBERTA FOREST HISTORY PHOTOGRAPHIC COLLECTION

Then began a rat hunt proper. I guarded the window to prevent him escaping again while the boys hunted and slashed at him with anything that was handy. One of them finally clobbered him with the broom. He opened the door, picked the rat up by the tail and was going to heave it out into the yard, but the hide slipped off the tail and the rat dropped at his feet. This nearly caused him to lose his supper as these pests have a most nauseating smell. All this caused quite a little excitement and the cabin was a shambles when it was all over.

An Indian boy called to know if the boys wished to buy any gloves or moccasins. He was wearing a gaudy buckskin coat and I think he could have sold several like it, but unfortunately the women had none made up. The boys placed quite a large order for moccasins and gloves, and the women were busy all night long, as we were leaving next morning.

We were having a real nasty spell of weather at that time and it was getting colder every day. However, by coming back towards Entrance on the Lower Trail we could use the ranger's cabins as they were now deserted. From Grande Cache to Muskeg River Cabin was a long journey, and it meant crossing and recrossing the Muskeg River several times. The ice was bare and the horses were afraid of it. We had a battle at every crossing to drive them in and break the ice, and their tails became solid chunks of ice. The saddles and packs were also covered with ice, which made packing and unpacking a miserable job.

To add to our misfortune, a young Englishman who had been an assistant ranger had burned up the firewood at each cabin. He had even split and used

nearly all the chopping block at one place. The code of the woods – and it is strictly adhered to – is that you leave as much wood as you find, or as one wag put it, a little more wouldn't hurt. I hate to think what would have happened to this young man if we had caught up with him, but he had quit and left by the time we got back to Entrance.

The weather improved and by the time we reached Headquarters at Entrance it was nice. We spread most of our equipment on the lawn and the saddle blankets were hung up. In a couple of days everything was ready to be stored in the cache until the next season.

I had a few days accumulated leave to my credit, so I went to Edmonton and caught a cold. That is something I could never understand. An individual, or a whole crew could live under canvas, like we did, wear wet clothes day after day, sweat or be almost frozen and never catch a cold. Then as soon as we went to the city, with all its comforts, we got real beauties.

Ernie Harrison and I had a few fires to put out that fall. The wind blew steadily for a week or more, drying the vegetation in the timber, and about this time of year it seems that hunters are not too careful with their cigarettes and matches. I refer to the local hunters around the settlement, as we never seemed to have trouble with regular hunting parties from outside in the charge of experienced guides.

Camp fires are sometimes critical to survival in the backcountry. In the absence of paper or other fire starters, experienced campers will often carve "feather sticks" such as these fire starters carved by Bob Stevenson.

BOB STEVENSON

Sketch of Entrance Store by Fred Tietge, c. 1940s.

FRED TIETGE COLLECTION, COURTESY OF STAN TIETGE

A Busy Winter in Entrance, 1925

The Canadian National Railway engineers went into conference and had decided to switch the track from one side of the Athabasca over to the abandoned Grand Trunk Pacific track on the other side. Their reason for the change was that the many high wooden trestles on this section were becoming dangerous and the cost of maintaining them was prohibitive. Filling in could not be considered either.

This was bad news for us as we were utterly dependent on the railway. The Jasper Highway had not reached that far as yet, and even when it did, it would be four miles away and we had no connecting road.

The railway bridge linking our community to the south side of the Athabasca River was to be dismantled, which meant we would be completely isolated unless the Department of Public Works could be prevailed on to give us a ferry. However, awaiting their pleasure was much like Mr. Churchill's famous words, "the mills of God grind slowly." However, Harry Davison, a multimillionaire who owned a dude ranch on our side of the river, solved the problem by purchasing the bridge.[24]

The new proprietor of our general store in Entrance, Tom Monaghan, started building a store on the other side of the river, so that meant we just had to have some sort of road out of there. Nothing could be done about it, though, until the steel was pulled, about two years later.

The CNR used the Grand Trunk Pacific grade for about ten miles east of Entrance from a point almost directly opposite our little hamlet. From there, a new grade was built upstream three miles to the new bridge site. During construction of this bridge, Entrance enjoyed a boom. The bridge men were making good money and they spent it freely. In the evenings they came down on speeders or on foot. They would play pool until midnight and then naturally they wanted something to eat. After eight p.m. meals jumped considerably in price, but they paid cheerfully.

Harry S. Davison

Harry S. Davison came from New York City to the Entrance area in 1928 to regain his health. He bought the Entrance Ranch from Stan Clark and, with his wife Anne, bought additional land. In 1930 they built a house on the Athabasca ranch, north of Hinton. When he died in 1961, his daughter Frankie inherited the Athabasca Ranch, and daughter Anne the Entrance Ranch. (Hart 1980). Mary Luger notes that the planks were not laid until the steel tracks were removed from the bridge in 1932 and that Davison may have bought the planks but not the bridge.

[24] Churchill was quoting "Retribution" by Henry Wadsworth Longfellow: "Though the mills of God grind slowly, Yet they grind exceeding small; Though with patience he stands waiting, With exactness grinds he all."

The Canadian National Railway

In the early 1900s, two railroads were built through the Hinton area – the Grand Trunk Pacific Railroad and the Canadian Northern Railroad, which passed through the hamlet of Entrance. Both soon went bankrupt and were taken over by the Canadian government. The new railroad, which became the Canadian National Railroad, chose the best parts of each route, linking them together into the one railway line that passes through the area today. By 1917, the initial consolidation was complete and the route along the east side of Brûlé Lake, built by the Grand Trunk Pacific, was abandoned – primarily because of the continual battle to keep the rails clear of wind-blown sand from the lakebed, which was and is mainly dry for much of the year. The main line crossed to the north side of the Athabasca River on the Canadian Northern Railroad bridge located east of Entrance, which then truly became the single "Entrance" to the Rockies. Then, because springs and slumps along the grade west of Entrance made it difficult to maintain, the CNR built a new bridge farther west. When the new bridge was completed in 1927, the railroad line through Old Entrance, where the Dominion Forestry Branch headquarters was located, was abandoned, as was the bridge over the Athabasca River. This bridge is now part of Highway 40.

Grand Trunk Pacific Railroad grade along Brûlé Lake near Park Gate Station, 1915. Note the drifting sand filling in the track and ties, a major problem that led to the abandonment of the Grand Trunk Pacific railway grade along the east side of Brule Lake.

M.P. BRIDGLAND 1915. PARKS CANADA COLLECTION

Rangers used various "speeders" to move around their districts. In this case, the unidentified worker is using an arm and foot powered Forestry Railroad Speeder, 1930s.

DOMINION FORESTRY BRANCH, ALBERTA FOREST HISTORY PHOTOGRAPHIC COLLECTION

Nearly every bridge of any size during the course of construction takes its toll of human lives and this one was no exception. Four men lost their lives when a section of the bridge collapsed and plunged into the river. This cast a gloom over our little community and the crew kept away from the poolroom for some time.

On this new piece of grade, there were several fairly sharp curves, and it was all through heavy brush, so Ernie Harrison and I decided to widen the right of way at those points. This would not only lessen the fire hazard, but would also provide better visibility for ourselves when we had occasion to use the Forestry gasoline speeder on the track. Anyone using a speeder on other than a straight track is under a constant strain, as when you round the next curve you could meet one of those huge juggernauts, a steam locomotive.

We used all the wood we cut, some for fence posts and the balance for firewood. It came in handy as everything could be hauled across the river on the ice. The limbs we piled and burned. This work together with the usual chores kept us busy for the better part of the winter.

The Tea Dance

As usual we had quite a few dances in the big log schoolhouse, which Mr. Woodley had built for the education of his large family. He also hired a teacher whose salary was paid by himself.

The Indians dearly loved dancing, and I looked in on what they termed a tea dance. They did not restrict it to tea either, and no holds were barred. Once, when a square dance had just finished, I asked the caller why he did his calling in English rather than in Cree. He told me that the Cree language is somewhat curtailed, and a great number of words have to do double duty with emphasis on inflection also. "Call in Cree no good," he said, "too much explain."

Opposite page: Long after it was finished as a Schoolhouse, the Woodley Schoolhouse at Entrance was used for social and other functions such as the October 11, 1952 wedding of Bill and Jackie Hanington.

COURTESY OF JACKIE HANINGTON

At one dance there was an Indian girl, Clara, who had the nickname of Step and a Half. This name had no doubt been hung on her on account of the very pronounced limp that she unfortunately had. I had been dancing a tag two-step and having been tagged I was without a partner. Clara was also without a partner and asked me, and to be honest about it I don't believe I ever had a better partner. The limp seemed to synchronize perfectly with the music. I asked her for another dance later on, but that was not a two-step. It was pure agony while it lasted and if Clara had been wearing spike heels instead of moccasins, I would no doubt have been crippled for life.

At another dance where a great deal of liquor had been consumed – if empty bottles mean anything – I was amused by one character wandering around in a daze. Finally he accosted me, and after a spell of hiccups, asked if I had seen a toque anywhere. I glanced across the room where I saw a toque under a bench where a lady was sitting. Drawing it to his attention I asked if it was his. He hiccupped that he thought it was, but seemed reluctant to retrieve it. I asked him what he was waiting for, and he said "I was just (hic) figuring that that dame (hic) might be (hic) a bit touchy about her (hic) legs." I resumed dancing and left him to his fate.

Murder at Fish Lake Flats

As time went on I got to know the Indians and their customs very well and found them on the whole a happy and carefree race, except when some of them became crazed with liquor. Once when they had congregated at Fish Lake Flats[25] (about eight miles from Entrance), before the annual pilgrimage to Lac Ste. Anne, the women had erected the teepees and everything was in readiness for their departure the next morning. Then the trouble started. Evidently a quantity of lemon extract had been brought into camp and a few of the men got very drunk. Most of the others had gone to bed. Suddenly, one man grabbed an axe, entered a teepee and smashed the skulls of two brothers, killing them outright. A third man died a few days later.

[25] Today known as Wanyandie Flats, these flats are just southwest of Jarvis Lake in Switzer Provincial Park.

The person or persons who supplied the extract was never found. Jealousy over the affections of a young woman was said to be the cause of the trouble. I was quite satisfied that the extract never came from the local store, as Mr. Monaghan knew his Indians.

An RCMP sergeant and a detective arrived to conduct an investigation. They had a Ford car shipped up by rail from Edmonton to Brûlé for the occasion, as it was possible to drive to the scene of the tragedy. I happened to be riding into Entrance from Winter Creek with that same crazy horse that furnished us with the bucking exhibition I mentioned before, when the Ford came round a bend in the road. Evidently that horse had never seen a car before, or else he had made up his mind to embarrass me, for he reared up on his hind legs, gave a stiff buck or two and headed for a thick clump of willows.

He seemed determined to get me out of the saddle no matter how, and I was just as determined to stay in it, especially in front of the Mountie. When the horse reached the willows he went into reverse, and that, let me tell you, is not only hard on the nerves but also on the clothes. Eventually I got him turned around and, spurring him unmercifully, got him almost up to the car as the fight had gone out of him. The Mountie had stopped the car, as it was too good a show to miss, murder or no murder.

The Woodley Brothers

The Woodley family represents an important chapter in the history of Hinton and Entrance. Roy Woodley and his younger brother Earl arrived at Hinton in 1911 working with Canadian Northern Railway survey crews. Roy was awarded the Entrance Post Office contract in 1914, operating from "Woodley Bros. General Merchants" building. According to family history, Roy originated the name "Entrance".

Roy T. Woodley took homestead on SW2-51-26-W5 at Entrance and lived there with his wife (Dora Webb who had arrived in 1912) and children until 1933, with a short departure to Vancouver in 1925. Roy held title to this land until 1948, relinquishing title to his brother Earl who in turn sold to Wilbert (Billy) Magee in 1955. It was subsequently purchased and is currently owned by Mary Luger, the daughter of Carl and Mildred Luger (Woodley) and granddaughter of Roy Woodley.

Earl Woodley married Edna Badgley and they settled at Red Pass B.C. acquiring Red Pass Hotel from Ralph Stanton-Veale in 1923. They established E.F. Woodley General Merchandise at Red Pass and ran their store and a successful mail order business as well as the hotel until retirement to Abbotsford B.C. in the 1940's.

Earl sold his share of Woodley Bros. General Merchant's business at Entrance to Roy in 1923. Roy then operated briefly as R.T. Woodley General Merchant.

On June 16, 1925 Thomas Monaghan bought Roy Woodley's business and leased the store buildings at Entrance (now Old Entrance). When the new railway bridge bypassed Old Entrance, Monaghan relocated the store and post office in 1927 to the south side of the river to the new building he built beside the GTPR Dyke railway station. Dyke station had been sitting idle for about ten years as the railway steel had been removed in 1917 from Chip Lake to Pocahontas for the war effort. Dyke station was renamed Entrance. (Courtesy of Mary Luger)

Earl and Edna Woodley 1922. Edna was the daughter of forest supervisor Ward Badgley.

ROY WOODLEY COLLECTION, COURTESY OF MARY LUGER

When we commenced work that spring of 1925, I really had a hand-picked crew, for we had an extensive program planned. All of them were young, husky and willing, and we accomplished all we set out to do.

Cadet troop at the headwaters of Little Berland River, Berland River Trail, 1940s.

FRED TIETGE COLLECTION, COURTESY STAN TIETGE

CHAPTER FIVE

A RENEWED TRAIL PROGRAM, 1925–1927

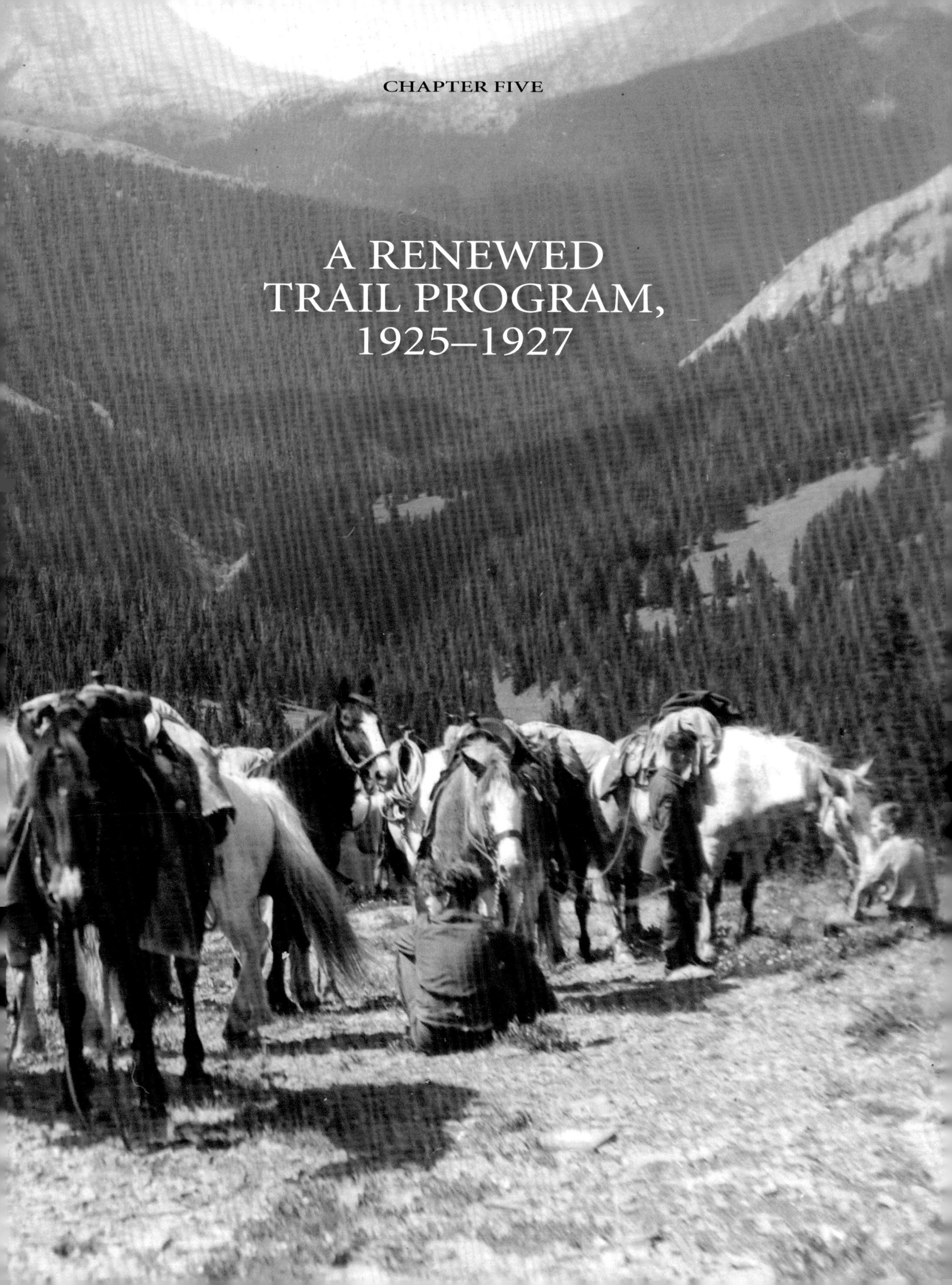

Berland River Trail

When we commenced work that spring of 1925, I really had a hand-picked crew, for we had an extensive program planned. All of them were young, husky and willing, and we accomplished all we set out to do. We cut around sixty miles of trail that season and did considerable maintenance too.

The first trail to be cut was down the Berland River to the fifteenth baseline, north of which was the Athabasca Forest boundary. For the first week we worked from the Little Berland Cabin. Towards the end of the week we were really walking farther than we should, but I could not locate a suitable campsite closer.

Our first move was to a camp at the junction of the Little and Big Berland rivers. Here was a camp that had everything a camp should have. The ground was level, wood was plentiful and for water a large spring bubbled out of the hillside. The water was crystal clear and cold as ice. Best of all, the fishing was supreme with Dolly Varden trout measuring up to thirty-three inches.

The first night in the new camp, I went to bed with the feeling that all was not well with the horses. Eventually I fell asleep, soon to be woken by the sound of bells fast approaching. The horses were evidently making a dash for it. With no time to dress, I grabbed a towel from a tent peg and jumped out of the tent brandishing it and shouting at the top of my voice. Paying no attention to where I was going, I stubbed my toe on a rock and fell directly in front of the lead horse. That was too much for him, and he wheeled around and the whole bunch galloped madly back up the trail. The next day our wrangler drove them up to Donald Flats and they caused us no further trouble.

The boys had a great time catching fish and the cook finally had to tell them to lay off as he could not possibly use what they brought in before it spoiled. They were crestfallen at the time and at the table gorged themselves with fish to try to create a shortage. Then they thought up a scheme. The spring water had over the years formed a sizeable pool. Into this pool they dumped the fish, which they brought from the river nearby. Next morning, though, horrors – every fish was dead, killed by suffocation I suppose, on account of the water not having had time to become properly aerated.

Undaunted they scooped up gravel and created another pool further downstream. This one was successful, so now they had quite an aquarium, or fattening pen as they termed it. In a few days this became overstocked and the fishermen had to content themselves with throwing the ones they caught back in the river again. Whenever the cook wanted fish he took a long-handled shovel and scooped them out at will. This worked well when the pool was so full the finny monsters could hardly get out of each other's way, but as their numbers decreased the survivors became pretty elusive. The cook enjoyed this sport himself and plied us with fish until everybody got tired of the sight of them.

An Innovative Cook

We did not have our good-natured cook of the previous season, as Don Empson had leased the restaurant at Entrance and had been joined there by his wife and small daughter from Edmonton. George, the new cook, had worked in mining camps and found camp cooking a bit more complicated. He told me a little wrinkle, which sounded good to me. Late one fall the company that had hired him was going into the Interior to do some development work, which would last most of the winter. The problem of vegetables, especially potatoes, kept him guessing for a time as dehydrated ones did not seem to be the answer. He peeled two sacks of potatoes, boiled and mashed them, froze them and packed the lot in cardboard cartons. Then all he had to do was to chop off a frozen chunk and warm it up. For bread he bought a huge quantity from the baker in town and froze that also. He had the men dig a hole under the cookhouse for storing food that was to be kept unfrozen, like canned milk, etc. He had thought of freezing the eggs also, but decided against it. Instead a couple of extra packers were hired to bring out a load of supplies, including two, thirty-dozen crates of eggs.

The eggs posed quite a problem for the packers, as they had to be brought into their tent every night and wrapped in blankets to prevent freezing. The gentlest horse in the outfit was entrusted with them. During the day, although the weather was cold, the heat from the horse's body kept the eggs from freezing. Then after all the trouble those boys took, disaster overtook them when they reached their destination. The cookhouse was built on a knoll and the steps leading to it were icy. Result: 30-dozen eggs smashed, a total casualty.

Camp Move to Donald Flats

All good things come to an end eventually, and we had to leave our pleasant camp. We moved two miles beyond Donald Flats and let the horses drift back there. Mr. Scott and his cattle had moved away from there, otherwise I doubt if our horses would have associated with the cattle.

The afternoon we moved, the sky was leaden and the horses jittery, always a sign of an impending storm. We worked like Trojans to get our camp shipshape before the storm hit, for if you pitch a tent on wet ground, it develops into a puddle in no time. We had the last guy rope tightened and the last piece of equipment under cover when there was a vivid flash of lightning accompanied by a terrific peal of thunder. It was followed by a regular deluge of rain, and in a few moments the hail started to pepper down. Apparently we caught just the edge of the storm for it let up shortly afterwards and we heard it gaining in intensity as it moved away.

After supper the stars came out and it turned cold, so we all went to bed to keep warm. We had had no time to cut wood for the sleeping tent stove and just barely enough for the cook to make breakfast.

Hunting camp at Donald Flats, 1986. The Berland River remains popular for canoeing, also for hunting parties such as this one travelling by canoe and inflatable raft.

BOB UDELL

Next morning I went ahead to locate and blaze trail. What a mess the storm had left in its wake. All the deciduous trees had been stripped of their foliage and the rotten wood lying on the ground had been battered to a pulp. The hail lay on the ground to a depth of four inches and was so tightly packed that I hardly made a dent when I walked on it. I sympathized with prairie farmers who are often subjected to storms like this.

As we were now working in a desolate area, we worked hard to finish the trail, which should have ended at the Forest boundary, and in the end, we overshot our mark and could not find the survey line. We had been cutting trail alongside the river and had reached a spot where the valley widened. There had been no recent fires that could have obliterated the line, so I finally decided that the bush in the valley, mostly willow and alder, must have grown over the baseline survey line and hidden it completely.

I was still not satisfied, so I climbed the bench and after a great deal of searching located the line. The surveyor had taken a long shot across the valley and had not bothered to cut the line in the valley bottom.

Back to the Mountains – Adams Creek and Kvass Trails

Our next project was the completion of the Adams Creek Trail. We commenced work at the crossing of the Big Berland River and, to avoid numerous crossings, kept to one side as much as possible. This was the route that Jim Mills and I had travelled going to the fire where he had his boots burned. We had a lot of hard work on this trail after we left the river, but we were within sight of the mountains and the boys were anxious to do some goat and sheep hunting as soon as the season opened, so they did not complain.

One evening we sat down to supper and attacked a huge meat pie that the cook had prepared. It smelled nice and really was delicious. Everybody was hungry and the whole pie was eaten. Now the cook was an expert at disguising such things as bully beef, bologna and other forms of canned meat, so I gave it no further thought. Afterwards he asked us if we knew what was in it and, after all had said how good it was, he told us it was porcupine. The boys seemed thrilled over it, but none of them had watched the Indians eating the practically raw flesh as I had.

The gypsies in Europe consider the porcupine, or hedgehog as it is called over there, a delicacy, but they cook it in a different way. First the entrails are removed, the cavity thoroughly washed and then it is covered with a thick layer of clay and buried in the glowing embers of a fire for a lengthy period. When it is considered to be cooked, the clay covering is peeled off and with it come the hide and quills.

Porcupines are clean animals and vegetarians, and if I had to choose between eating frog legs or porcupine, under protest I should certainly choose the latter.

Looking west across the Berland River to the Mountains. Adams Creek enters the Berland River from the right, 2006

BRIAN CARNELL PHOTO, FOOTHILLS RESEARCH INSTITUTE COLLECTION

The Adams Creek Trail being finally completed, we backtracked to the Mountain Trail and reached Rock Lake on a Saturday evening. We did not travel on Sunday and the boys certainly enjoyed the stopover. They swam, climbed the mountain and fished in the evening, so we had lake trout for breakfast on Monday morning. We made the journey north in easy stages as our horses were all carrying heavy loads, our packer having just brought out a fresh supply of food from Entrance.

The first stop was at Eagles Nest and I was disappointed to see that nothing more had been done to the cabin since Sam and I left. It was deteriorating fast as campers had used up the bunks, the table and everything else for firewood. The yard was littered with cartons, tin cans and bottles. Truly a disgusting sight.

As we went through Eagles Nest Pass, the boys took quite a few pictures. When we emerged into Rock Creek Valley, the meadowlarks were there as usual to serenade us. I have travelled the length and breadth of the Athabasca Forest Reserve, but have failed to see or hear the larks anywhere except in this valley.

One of the boys remarked that this wasn't work – it was a Cook's guided tour.[26] We next camped at Rock Creek Summit or Mile 58 from Entrance. This was a lovely campsite, and the bunch grass was plentiful so the horses never tried to stray. It was here that the boys had their first look at caribou. In a shady basin

[26] A Cook's tour is a quick tour or survey, with attention only to the main features, supposedly named after Thomas Cook (1808–1892) who founded the travel agency that still bears his name.

Pack train along upper Sulphur River, 1914 Fay Expedition.

FAY EXPEDITION, JASPER YELLOWHEAD MUSEUM AND ARCHIVES JYMA84.87.49

on a mountain a few miles from camp there was still snow and here the caribou were gambolling and having mock battles. We watched their antics through binoculars and would have liked to get a closer view, but it was too late in the day. A few years later we built a cabin here as it broke the journey nicely when going to Big Grave Flats.

A few bridges claimed our attention in the Little Grave District. I showed the boys the cabin where I spent the memorable night with the mice. Everyone joined in a laugh at my expense, and one of them remarked that I was fortunate to be alive when you considered what had happened to the Bishop of Bingen in his Mouse Tower on the Rhine. According to legend, he was destroyed by mice, and after my experience I could see how it could happen.

The Kvass Creek Trail was our next project. We began cutting on Big Grave Flats close to where that creek flows into the Sulphur River. No great difficulty was experienced until we reached Kvass Summit, and then our troubles started in earnest.

We hoped that by camping in the pass, which was narrow at this spot, we could prevent the horses from going back to Grave Flats, a distance of eleven miles. Our hopes on that score were shattered the first night. We had been delayed in leaving Grave Flats as the horses had split into two bunches, and one horse fell into a deep hole alongside the trail and got the pack wedged. After a lot of work we got him mobile. It was almost dark when we reached our campsite.

Weather Challenges

By this time a terrific wind had sprung up, which didn't improve matters, but at last we got both tents up and our firewood sawed with the aid of a flashlight. By the time the cook had supper on the table, it was bedtime and as everyone was tired we soon fell asleep. The wind had increased so we did not hear the horses pulling out, but in the morning they were gone. One could hardly blame them as we were close to timberline and, although vegetation was rank, it was mostly weeds and very little grass.

Sometime during the night I was awakened by a terrific crash and was at a loss to tell what had happened until I shone my flashlight on the scene. The wind had blown down an old dry snag, which in turn had smashed the ridgepole of our tent. My first concern was to find out if anyone was hurt, but all hands answered saying they were OK. Struggling out from under the canvas and getting our clothes on took some doing, but we finally made it and managed to get another ridge pole by the light of the moon. In all of our other camps I had made doubly sure that all dry snags likely to fall on the tents were taken down, but on this occasion lack of time had prevented us from doing so.

We had only worked a few days out of that camp when it started to snow. Thinking it might only last a day I told the boys there would be no work that day and we would work on Sunday to make up for it. With lots of magazines to read and a good fire we suffered no hardship. The snow fell for two more days, but we could not afford to wait, so we worked. The weather was not too cold, but every time you hit a tree you got a shower of wet snow down the back of your neck. I fared worst of all, as I hit more trees than the others while blazing trail and to make matters worse I fell in the creek a couple of times. Then the wind started up again and blew most of the snow off the trees so we could make better headway.

The Kvass District is very good for game, so now that the trail work was progressing favourably and there were a few days due me, I decided to take a hunt. The weather was nice around the summit, but I knew that it would be cold on the mountain peaks and my heavy mackinaw coat would be burdensome for climbing. Instead, I wore a buckskin coat that one of the boys offered me as it would be light and yet warm. When I reached the top it was blowing with terrific force and the coat inflicted punishment beyond belief. The trouble lay in the fringe and this slapped my ears and face bringing tears to my eyes. Particles of ice and rock blowing around added to my discomfort. I had a hard time keeping my footing as I dashed from shelter to shelter.

Reaching the lee of a rock bluff, I looked the country over and saw no sign

Ranger Gordon Watt from Entrance crossing the Smoky River in Stan Clark's Boat, 1948.

GORDON WATT PHOTOS, ALBERTA FOREST HISTORY PHOTOGRAPHIC COLLECTION

of sheep. They probably had more sense than I and were sheltered somewhere. I did see a few goats on a distant peak but I was not interested in them, so I made my way back to camp. Going down to the Smoky River in that tortuous valley, I avoided as many crossings of the creek as possible, but there was little choice. The scenery was wonderful, but apart from that the trail was a headache. The trapper in that district had built a queer-looking cabin. He had dug down about three feet or until he struck rock. Then from the ground level he built four-foot log walls. A pole roof covered with the soil he had excavated completed the edifice. In a few years' time the roof was completely covered with a growth of wild black currants.

What a relief it was when we were able to leave that valley and go the rest of the way to the Smoky on higher ground. There were stretches of that trail where the sun never penetrated and the din from that energetic little creek got on a man's nerves. When we finally reached the Smoky valley it was just like emerging into another world.

We used Stan Clark's boat to cross the Smoky. The crossing was at a huge whirlpool and caused more than one nervous person to gasp. I got so used to it that I have often crossed after dark. The campsite there was not all it might have been, as the fine mica sand from the river seemed to get into everything. It was also infested with mice. I made up my mind that a cabin was much needed here, and eventually I had my wish.

As I mentioned before, Stan Clark had a boat there but another was needed so that one could remain on each side. This particular trail had priority, so another boat was packed in the next year.

A Busy Fall and Winter

Our season was once again drawing to a close and we moved camp down river from Big Grave Flats to do a little repair work before heading back to Entrance. One evening while coming home from work, a goat was sighted on the skyline of a mountain about three miles from camp. Grabbing a quick snack at the cook tent, three of the boys started after it. Two had rifles and the third went along for the excitement. They lost no time getting up there and I imagine it was due to their being out of breath that they missed as many shots as they did. Probably there was a bit of buck fever besides. We watched them through binoculars and when the shooting stopped they continued darting here and there until we finally lost sight of them altogether.

When it got dark I began to worry, as mountain climbing can be dangerous in daytime let alone at night. We built a huge fire out in the open to guide them back down and the cook kept the coffee hot. At long last they arrived and what a mess they were. Their clothes were torn, they were covered with blood and hair from the goat and I thought how fortunate it was that no devil clubs grew in the Alberta mountains as they do in British Columbia, as they might have been a sadder sight.

The explanation was simple. They had run out of ammunition, but the goat was wounded and like all his kind, he kept them at bay with his back to the wall. They finally stoned him to death, which took a lot of doing. Rocks of a suitable size for throwing were hard to find, but they finally worked out a scheme that brought results. Two of the hunters peppered him with whatever was available, while the third climbed above and dislodged a huge rock, which made a direct hit. He was still trying to get back on his feet when they reached him and cut his throat.

The Bargain Buggy

Shortly after this, we started back to Entrance. A few miles from the end of the journey, two of us stopped to adjust a pack that had slipped. We were going at a trot to catch up to the outfit when I noticed that each horse ahead would reach a certain spot on the trail, look down the hill, shy and make a wild dash past.

We soon located the cause of the commotion. Here was our friend Johnnie, the same man who had remarked on the creek not having run dry for twenty years. He was crouched over the remains of what once had been a horse buggy. By the huge grin on his face – standard equipment for Johnnie – he did not appear to be greatly worried, and I asked him what had happened. He explained that he was tired of packing horses and figured it would be much easier to move his worldly goods on wheels. Mr. Woodley had an old buggy that had been standing out in the weather for years and was beginning to come apart at the seams. After a great deal of coaxing on Johnnie's part, they succeeded in making a deal at the staggering price of five dollars.

The people of Entrance had turned out to watch his departure. Nothing untoward happened until he reached a steep little hill where a brace in the body of the buggy became dislodged and got mixed up in the spokes of the wheel. He now had a noisemaker superior to anything used to usher in the New Year, and his fiery mustangs broke into a mad gallop, jettisoning the contents of the buggy along the road, Johnnie included.

This was more than the spokes could take, and one wheel started to disintegrate with the rim flying off at a tangent into the bush. The main body turned over at the spot where we found Johnnie, and the team continued triumphantly toward home bearing only the buggy tongue and the whippletrees, which no doubt lent speed to their heels. To listen to Johnnie telling it was too funny for words, and we laughed all the rest of the journey. The saying that some people laugh at other's misfortunes seemed very apt on this occasion.

When all equipment was put away for another season, road building was the order of the day. Soon enough, the railroad through our community would be abandoned in favour of the route on the other side of the river, where a new community (also to be called Entrance) was growing. Mr. Monaghan had built his new store across the river and intended to move in when the river ice was safe enough. We could do nothing yet towards making a road on the portion of the track to be abandoned, as it was still being used.

The Athabasca Bridge

Mr. Davison, the owner of the dude ranch, had built a road from the railway bridge up the hill to his own property. We had an existing road that was passable, so we put a road from it through our east pasture to connect with the dude ranch road, and planked the railway bridge.

At the east end of the bridge, there was a deep rock cut where a watchman patrolled the track at night. He built a rough shack out of odds and ends and, during the summer months, had sweet peas trained on the walls, which transformed his otherwise rough looking shack into a beauty spot. At the far end of the rock cut, we continued our road building to give access to the new village-to-be, branching off to connect with the Jasper Highway then under construction.

To cut about a mile off the distance between our community and the new Entrance, a long ladder was placed against the cliff and pedestrians used it extensively. Concerning this ladder, there was an incident that might have been a serious accident, to an Indian lad. This young man had taken on more liquor than was good for him and was taking the short cut. The rock ledge at the top of the ladder had a slight downward slope to it and on this occasion it had a light covering of snow. The lad was still facing forward when he stepped on the ladder with his moccasins, and at that moment his feet went from under him and he coasted over the edge. He was wearing a heavy mackinaw coat, unbuttoned, and the ladder rails slipped up the back of it to his armpits, leaving him suspended.

The ladder at the south end of the Athabasca Bridge with the Woodley family and car, returning home from school 1933. Planks had been added to the abandoned railway bridge to carry people, horses, cars and trucks.

ROY WOODLEY COLLECTION, COURTESY OF MARY LUGER

The rungs of the ladder were slippery also, affording no grip for slick moccasins. He sobered up very quickly and called for help.

Mange and the Bounty Program

Tom Burrows was granted leave of absence that winter to go to England on private business, and while he was away Mr. Edgar,[27] supervisor of the Brazeau Forest Reserve, handled the mail and visited us through the winter to discuss future plans.

A fly shelter for the horses was badly needed at Moberly Creek, so it was decided that Bill Douglas and I would cut and haul the logs for it as soon as I had the baled hay all hauled. We bought our supplies and loaded up with enough grain for the team and got as far as Winter Creek as the sleighing was poor and snow had to be thrown on bare spots.

The next morning I was leading the team out of the barn when Bill spied a coyote trotting along the sleigh road and coming in our direction. He went to the cabin for his rifle and I put the team back in the barn, closed the door and watched the proceedings through a knothole. The hay baler was parked close to the road and Bill hid behind it.

[27] Frederick George Edgar started as a forest ranger in 1912 and became forest supervisor on the Athabasca Forest from 1933 to 1935. He worked as forest supervisor on other forests until he retired in 1951 after 39 years of service.

On came the coyote, evidently not suspecting any danger, and when he was opposite the hay baler Bill pulled the trigger. That coyote must have got the surprise of his life for he made a tremendous leap and vanished without a scratch. It was amusing to see the look of consternation on Bill's face, but when he looked at the hindsight of his rifle he saw why he had missed. When he had last used the rifle he had raised the hindsight for a distant target, and in his hurry he had forgotten to lower it again.

At that time a great many wolves and coyotes suffered from mange, and here's why. The losses to predators sustained by both cattle and sheep men had reached such alarming proportions that they decided that something had to be done about it. Bounties were increased and organized hunts were staged, but it had done very little to relieve the situation. Poisoning was too dangerous and besides, the killers were suspicious of any kind of bait, especially when they could have all the live meat they wanted. The idea of injecting mange into animals caught in traps and set free was adopted with fair success south of the border and the disease soon spread to Canada. A mangy coyote will invariably seek the shelter of a building when he feels he is about to die, although he will give all buildings a wide berth when he is hale and hearty. I have cremated numbers of dead and dying coyotes taken from the hay sheds of both Winter Creek and Wildhay River ranger stations.

Winter Creek Cabin, 1947.

MARK TRUXLER COLLECTION, COURTESY OF JACKIE HANINGTON

As the bounty on coyotes was five dollars over what the pelt would fetch in good condition, I sometimes hunted them. One day I noticed one in the hay meadow and figured that he was hunting mice. As the distance was great and I could get no closer without him seeing me, I decided to try a shot anyway and when he dropped I certainly was very much surprised. When I reached him he was to all appearances dead and what a revolting sight it was. The hair was practically all off his body and tail and he was a mass of scabs and sores. I thought to myself that even if those animals were pests this was a most inhumane method of exterminating them.

As every hunter likes to know where his shot was hit, I turned the body round and round with the rifle barrel but could find no trace of blood or a bullet hole, so came to the conclusion that the concussion of the bullet passing close to him in his weakened condition had finished him, or he could have been anemic and no blood issued from the wound. I gathered some brush and old hay and cremated him where he lay.

The coyote that Bill shot at showed no sign of mange.

I returned to Winter Creek the following day for a few bales of hay for horse feed. We had just started felling the timber for the building when it began to snow – and it did not stop until sixteen inches had fallen. This made our work very difficult, but to quit now would have meant postponing the building of the shed for another year, so we carried on. The job took longer than we had figured and the result was that we were not only short of horse food but mighty shy on food ourselves. We could not wait for a chinook wind to take the snow away, so we started out back to Entrance, and what a trip it turned out to be. Along one stretch of the road where the snow had drifted we had to go ahead of the team and tramp a trail, otherwise the horses could not have made it. The lights of Entrance were a welcome sight to us, as it was late and we were hungry and dog-tired.

I was sorry to learn that Louie Holm was quitting and going farming. He died a few years later. He had been feeding the horses at Wildhay River, so I took over for the rest of the season. I enjoyed it out there and could visit my friend Harry Phillips quite often. He was engaged in commercial fishing and doing very well at it.

Wildhay–Moberly, À La Pêche and Muskeg Trails, 1926

When the new season started, my crew this year was smaller than usual, but we did not have many projects. Blackie was with us and this was a big help as he was very capable and took some of the responsibility off my shoulders. We commenced cutting on the Wildhay-Moberly Trail and everything went fine until I started looking for a suitable ford. I liked to make all the trails as short as possible, so I did not feel like going too far out of the way to find a ford. The crossing I finally chose was not as good as I would have liked, but I consoled myself with the thought that there were lots worse and that anyway this trail would not be used much.

Telephone Line to Muskeg

When this trail was finished we moved to the Muskeg ranger station. As the ranger was away we saved ourselves the trouble of pitching the cook tent by using the cabin instead. This was the end of the telephone line and was sixty-two miles from Entrance.

The day we arrived, the boys were curious to know how some of the telephone equipment had been brought there. They figured that it was too unwieldy to have been packed on horses. For instance, there was a heavy steel bar about seven feet long, a shovel with a handle eight feet long, and two parts, known as spoons, with handles the same length. Those spoons are far more like dippers and, as the name implies, are used for spooning or dipping the soil out of deep holes. Finally, one of the boys asked me how those implements got there and when I told him by manpower, he gave me a withering look, figuring no doubt I was stringing him along. I explained that the line had been put in by a gang of Swedes under contract. The government had furnished the pack train and a packer to transport their food and camp equipment. The men had walked the whole distance hanging the wire on trees or digging holes and setting poles where necessary, so they must have carried the tools with them all the time.

Murder at Pierre Grey's Lakes

Pierre Grey's Lakes was the name given to a chain of small lakes not far from the Muskeg Cabin. On one of these lakes, Blackie and I built a raft and went fishing for lake trout in the evenings or on Sundays. May 24 being a holiday, I took a lunch and went alone to fish. It was a beautiful day and as the song goes I was "drifting and dreaming." I could see an occasional fish, but they did not appear to be interested in my bait.

One seemed a bit curious about the raft as he circled it slowly, then swam underneath from side to side. I got down on my stomach to watch him better. While I was gazing down through the pellucid depths, something white on the bottom caught my eye. There was a slight shimmer on the surface which prevented me from seeing the object too clearly, but I was convinced in my mind that it was a skeleton – whether man or animal I could not decide – but I imagined that I could see the rib structure. It was too small to be a moose and too big to be a wolf. That meant it must be either a mule deer or a human.

Suddenly a tale that I had heard – and to which I had not given too much credence – came to mind. A trading post had been built and stocked with merchandise by a man called Pierre or Peter Grey, and it was said that he worked up a good trade. One year during the winter, a band of Indians came to the post to sell their furs. Someone produced liquor and a drunken brawl followed with one man being killed. This was kept secret for many years and the trading post was abandoned. Shortly afterwards, a young Cree who had been studying for the priesthood at a Catholic college suddenly quit and returned to his tribe, and turned to trapping. Rumour had it that he was the brother of the murderer and a stigma had been placed upon him and his family so that he could no longer continue with the work of the church.

I lost no time in paddling the raft to shore so that I might search for signs of a building, but found none as a forest fire had swept through that part of the country within the last thirty or forty years and new growth had sprung up.

I never reported my findings to the authorities as nothing could be gained at that late date and if the remains were human, why not let him rest in peace. But I lost all interest in fish and fishing in that lake.

Connecting to the Mountain Trail

Our next move was a long one as there were no suitable camps between Muskeg Cabin and À La Pêche Lake. At the latter point there was now a cabin and, as it was unoccupied, we moved to it. The country around the lake was swampy and favourable to moose and we often saw them swimming in the lake, no doubt to escape the flies. As already mentioned, fishing was excellent in À La Pêche Lake and there is wonderful fishing also in À La Pêche Creek, which is not very long and flows into the Muskeg River. Fishing was also good in the river itself.

One day when I was trying my luck I came to a likely looking spot. A logjam

Pierre Grey

Pierre Grey was born Pierre Gris at Lac Ste. Anne in 1846. His mother Suzanne was from the Calliou clan; his father, Joseph Grey, was a Métis born in the Jasper area in 1809. Pierre Grey was a successful free trader (not associated with a major fur company) and ran a trading post during the winters at Pierre Grey's Lakes near Grande Cache in the late 1800s. Summers he spent at St. Albert or his homestead at Isle Lake, where he and his wife Marie (Delorme) generously took in a number of homeless children. Pierre Grey and his wife died in the great flu epidemic of 1918; he is buried in the cemetery at Lac Ste. Anne.

Source: Richard Wuorinen, *A History of Grande Cache*, Grande Cache Historical Society, 1997. pp. 39,40.

Pierre Grey Lakes, 2006.

BRIAN CARNELL PHOTO, FOOTHILLS RESEARCH INSTITUTE COLLECTION

had dammed the river and created a pool. Not having been favoured with a bite, I decided to try the other side and started walking cautiously on a long slim log. I hadn't bothered to wind in my line, letting the bait drag along behind me. When I was about half way across, a big fellow struck and I almost got pulled in. I did not relish the idea of taking a ducking so I had to do a balancing act and play the fish at the same time. Fortunately the Dolly Varden is not a fighter like the rainbow or the steelhead and after a few desperate lunges he decided to call it a day. It certainly was a thrill while it lasted.

When we had the Muskeg River Trail cut as far as À La Pêche Creek, we abandoned it for the time being as I wanted to avoid back tracking. By cutting the À La Pêche Trail, which would connect with the Mountain Trail at Cowlick Creek, we would have to backtrack only seven miles instead of the twenty miles if we had built the Muskeg River and Rocky Pass Trails first. It would also mean that we would be twenty miles closer to town at the end of the season. By having a smaller crew that year, we did not accomplish as much as in former years but it gave the boys a full season's employment.

When the À La Pêche Lake Trail was finished, we resumed work where we had left off on the Muskeg River Trail. Once, while moving camp where the trail came close to the river at a narrow canyon, I surprised a golden eagle. He had just caught a fair-sized trout and had risen only about twenty feet in the air when he dropped it. We wondered if he would return and search for it after we had passed.

Cow moose and calf.

ISTOCKPHOTO

This was very good game country and seldom a day passed but we would see either moose or deer. The young moose learn their lessons the hard way. One day I saw a cow and her calf heading for the river. The calf could not have been more than a few days old as it was still wobbly on its legs. The mother entered the water at one of the deepest spots in the river and just above the rapids. Moose are very powerful animals and this one was no exception as she never swam a stroke and the water was almost up to her shoulders. The little fellow had no choice left but to follow and as he entered the water just above the rapids he was quickly swept downstream. Upon reaching the other side the mother looked round for her calf and, not seeing him, she became quite excited and commenced a sort of blatting call. Finally she saw her calf a long way downstream, bobbing like a chip on water but its head was still above the water. I have an idea that it eventually drowned.

Blackie and the Unwelcome Visitor

We had moved camp and only managed to get the cook tent up when darkness overtook us. After supper we covered everything up and each man made his bed under a spruce. I was tired and soon was asleep, only to be awakened by a rifle shot that sounded as if it were right at my ear. I have been awakened by various sounds, but nothing can come up to a shot for scaring the wits out of a person. I was in a cold sweat, and matters didn't improve when I heard one of the boys say, "Did you kill him, Blackie?"

"Well if I didn't I sure must have scared the hell out of him."

Ye gods, I thought, surely Blackie didn't commit murder. I coughed loudly to attract his attention in case he might mistake me for the fellow he was trying to do in.

However, instead of being a murderer, Blackie turned out to be a hero for he had shot a porcupine, which could have destroyed much of our saddlery. Porcupines crave salt and grease. Ernie Harrison told me about one time his partner had a pair of practically new Leckie logger shoes ruined by one. At the time they were using an Indian teepee instead of a tent and both being sound sleepers they heard nothing.

This night Blackie had awakened and heard something chewing close by. He shone his flashlight on the spot, and the porcupine made for the closest tree and started to climb it. Blackie woke one of the boys and had him shine his flashlight so he could get his rifle sights on the animal. He killed him all right, but he did not fall out of the tree as he was caught in the crotch. He had been chewing on a ham bone, which the cook had tossed out after supper.

Hardships on the Rocky Pass Trail

As we would soon have the Muskeg River Trail finished, I travelled up almost to the headwaters where we would begin on the Rocky Pass Trail. At the timberline a wonderful sight met my eyes – I was now at the end of the pass, or the beginning, depending which way you looked at it. It would be the beginning for our next project. It was a narrow alpine valley with flowers in bloom everywhere.

About half a mile further on, the valley was blocked by a great mass of rock, which at one time had been the top of a nearby mountain. Somebody had moved rocks aside here and there and made a sort of trail, which we improved as best we could, but it would have needed lots of dynamite and proper tools to make a good trail of it.

We finished most of the rock work and intended to move camp the following day, but during that night it started snowing and blowing and although we were well sheltered, we feared our tents would be ripped to pieces. For the next three days all we did was to keep our fuel supply replenished for ourselves and the cook, sleep and play cards. This got tiresome and we were glad to wake up to a lovely bright morning on the fourth day. It took a lot of preparation to get started – everything, including the horses' backs, was covered in snow and it was almost like ice. Fortunately, the horses had not wandered off, for we surely would have been in a fix with no snowshoes. We put their saddle blankets on, and after about an hour their backs dried out enough to put the saddles on.

Had I not intended to go only a short distance before establishing the next camp, I would have stayed put and waited for a chinook wind to start up and take away most of the snow, as it was still a little too early for snow to stay on the ground.

At last we got started, and what a journey it turned out to be. We took turns tramping down the snow ahead of the horses and eventually reached the pass. A long stretch of the trail was just a jumble of rocks and now the snow had drifted everything level. How we ever got through it without a man or horse getting hurt is more than I can tell. I took the lead going through the rocks, but as far as following the trail was concerned I'm sure I was off it more than I was on it. Time and again I had to be helped out of a hole into which I had fallen.

With the passage of several horses, the track became slippery and the ones in the rear would slide and fall. To make matters worse, our lash ropes had been frozen when we packed up and the heat of the sun, together with the heat from the horses' bodies, thawed them out causing the packs to become loose and slip. This had to be attended to right away.

Years afterwards, Bing Crosby wrote an article in the *Saturday Evening Post* that reminded me of that trip. He took time off from his Jasper Park Golf Tournament to hunt in this district and he claimed that he had personally slid down most of the mountains on the seat of his pants. We all did plenty of sliding that day, as did our horses.

Over a snowy pass, 1940s.

MARK TRUXLER COLLECTION, COURTESY OF JACKIE HANINGTON

I had a landmark in mind that day, a small lake just at the far end of this jumble of rock and it was indeed a welcome sight when we spotted it. Only a thin sheet of ice had formed on it and it showed up green amongst the drifts.

Snowblind

None of our party had coloured snow glasses, so we all suffered from snow blindness caused by the glare of the sun on the snow. In some of the stories of the north I have read, it is always during a blizzard that someone is stricken with snow blindness. A person who has experienced it knows better. It is quite painful while it lasts. Your eyes water and feel like they have sand in them.

When we reached our new campsite we had our work cut out for us. First we had to clear a place to unpack and then clear for the cook and bunk tents, pitch

Ernie Harrison breaking trail in deep snow, 1920s.

ERNEST HARRISON COLLECTION, COURTESY OF LUCY MCRAE

them and rustle wood. We were dog tired and when we finished getting things in order the next day, we only attended to the necessary chores and kept out of the sun as much as possible on account of our eyes. Bathing them in cold tea seemed to soothe them wonderfully and in a few days they were all right. There's something about high altitude snow or perhaps it is the sun's rays that affects a person more so than at lower altitudes. A few years later, two of us travelled through this same pass towards the end of June and, although the hardy mountain flowers were commencing to bloom, we still found quite a few snow patches. The next day our faces were not tanned but burned red and our lips were chapped and bleeding.

Our troubles were pretty well over as the rest of the going through the pass was not too bad. As I had been hoping, a chinook wind started to blow, and in no time at all most of the snow was gone and we had a lovely spell of Indian summer. I found two peculiar trees on that trail. One was about ten inches at the butt and at six feet it had tapered to seven inches and formed a complete loop. The other tree had begun as two separate trees two feet apart and when they were about ten or twelve feet high, they had come together and continued as one tree from there on. The twins had no doubt started from the same root.

We were in good Rocky Mountain sheep country and one of the boys was successful in bagging one. Blueberries and wild gooseberries were also to be found there, and all this was a boon to the cook and much enjoyed.

Opposite page: Stan Kitchen and Bing Crosby with trophy sheep, 1944. Frank Moberly partially visible behind Stan Kitchen.

MAC ELDER AND THE WILLMORE WILDERNESS FOUNDATION

Bing Crosby (1903 – 1977)

Bing Crosby the noted singer (crooner) of Hollywood fame, was also an avid golfer and sportsman. He first visited Jasper in1946 to film *The Emperor Waltz*, at which time he enjoyed playing the golf course at Jasper Park Lodge. He became a frequent visitor and won the Totem Trophy in 1947. He also hunted bighorn sheep in the Athabasca Forest north of Jasper Park in 1947, guided by Stan Kitchen with Frank Moberly and Frank Joachim.

(Feddema-Leonard 2007).

From Rail to Road, 1927–28

The Canadian National Railway had now abandoned the section of track on our side of the Athabasca River and had pulled up the steel. We now got busy and dragged the ties out and burned them. In the meantime, Tom Burrows had purchased a horse-drawn road grader for the Forestry Department, and we made good use of it and soon had a very good wagon road over to New Entrance. From there we started a road to connect with the Jasper Highway, which was now being built.

There was no appropriation allotted for this project, so we had to do it in our spare time, but everyone was anxious to see it completed as most of us had visions of owning cars in the not too distant future.

The only way we could get out of Entrance at that time was by train or saddle horse. There were some terrific hills there and when I finally did buy a car, the agent gave me a few driving lessons and told me that anyone learning to drive in a country like that need have no fear of driving any place else.

One carload of prairie people had the scare of their lives on a visit to Entrance after the road was completed. I have mentioned the rock cut at the end of the bridge. This was where the road left the railroad grade by a vicious hill running down at right angles to the grade. The brakes on this car were faulty and they were on the hill before they were aware of it. They just managed to negotiate the curve going at terrific speed and the driver had his hands full dodging the loose rocks in the cut, and before they had lost momentum they were on the bridge. This presented another nightmare as it was about a hundred feet above the swift flowing Athabasca River and had as yet no guardrail. It certainly spoiled their vacation, and the driver vowed he would not recross that bridge. I was told that the whole party walked back over it and one of their friends drove the car.

The hill on the Entrance, or north side of the bridge leading up to the dude ranch was also narrow and very steep. About that time, freewheeling had been installed in some models of cars. A number of the ranch hands were going to Entrance on this occasion and had borrowed one of the freewheelers. Whether the driver was not too familiar with the operation of this car or not I don't know, but he got started down this hill freewheeling and could not get the machine in gear. One of the girls in the party became almost hysterical and commenced shrieking at the driver to let her out. The driver as well as the others were no doubt wishing that they also could bail out. Fortunately no one was hurt when the car hit the steel gate at the foot of the hill, which is always kept shut on account of the stock.

I spent most of the winter looking after the horses. Ernie Harrison and I cut and hauled firewood after the snow came, as road building was out of the question after freeze-up. Each weekend, I took the team out with me to Winter Creek and brought in a load of baled hay. Before I left Entrance, I would toss out enough hay to last the loose horses in the community over the weekend.

Winter Creek Cabin and forestry staff, 1940s.

JACK GLEN PHOTOS, ALBERTA FOREST HISTORY PHOTOGRAPHIC COLLECTION

Canadian Northern Railway station agent John Christie with his wife Rita and Mrs. Ernie (Blanche) Harrison, seated at Entrance Station, c. 1925.

ERNEST HARRISON COLLECTION, COURTESY OF LUCY MCRAE

The CNR agent, Mr. Christie, and I were great friends as we were both radio enthusiasts and I spent many pleasant evenings with him. He was a bit lonely as his wife and family lived in Edmonton. I used to marvel at how he could be listening to a radio program or carrying on a conversation with me and at the same time be listening to the messages coming over the wire in Morse code. He would always make a cup of tea before I left, as coffee at that time was not as popular is it is now and coffee breaks were unheard of. I was saddened to learn of his death a few years after he had been transferred from Entrance.

When out at Winter Creek and Wildhay River, I visited Harry Phillips and our conversation was mostly about hunting and fishing. All in all, I put in a very enjoyable winter. The coming season would be the last of our trail cutting, as after the Smoky River and Sheep Creek Trails were cut we would have enough secondary trails to provide access to most of the Forest Reserve. Two lookout stations, Athabasca and Moberly Creek, with Berland River in the blueprint stage, would be sufficient coverage for fire protection, we hoped.

Building the Sheep Creek Trail, 1927

Spring was late in coming in 1927 and unfortunately we ran short of hay, which resulted in the horses being in a very weakened condition. Our first permanent camp would be at the Smoky River with the only hope of a rest being a Sunday or a very stormy day. Travelling time cut into our working time badly as we could average only around fifteen miles a day, limited by suitable horse feed and camp sites. We now had our own Forestry boat at the Smoky River crossing, which, with Stan Clark's boat, simplified the matter of moving men and equipment across. I had two men who were really good boatmen, but the remainder of the crew were hopeless and afraid of the crossing. When all equipment was put across we had quite a job getting the horses to swim, but when one of them got the idea the others followed like a bunch of sheep.

Building a Food Cache

As soon as we had camp shipshape, we built a food cache about ten feet off the ground. Our horses were heavily packed to get to this point, and it seemed foolish to pack all our food to Sheep Creek Pass, which was at the Interprovincial Boundary, only to pack part of it back again. We found three trees spaced about the right distance apart, but we could not find a combination of four so dug a hole and substituted a post. Using a makeshift ladder, we then sawed the tops off the trees, spiked cleats all around for the floor to rest on, and built a log wall about two feet high. For the roof we stretched a canvas over a ridgepole and our cache was completed except for peeling the posts and nailing tin part way up. This was necessary to keep black bears and other animals from climbing up and robbing the cache. Squirrels, too, can do a lot of damage, so we cut down all trees within jumping distance of it.

Building a cache to keep food and supplies away from bears, 1940s.

MARK TRUXLER COLLECTION, COURTESY OF JACKIE HANINGTON

First camp at Clark's Crossing on the Smoky. Jack Glen is second from the right, c. 1920s

JACK GLEN PHOTOS, ALBERTA FOREST HISTORY PHOTOGRAPHIC COLLECTION

Our packer made the round trip to Entrance at the end of each month, taking in my diary and report, and fetching back mail and such articles as canned milk, bacon and eggs. To save making two trips a day by boat across the Big Smoky, I decided to camp on the Muddywater River, which had its confluence with the Smoky not far from there. We had to cross the Muddy also, so I selected a spot where a small island in the centre of the river could be reached by falling trees from either side. Some people have a tendency to become dizzy when crossing water, so I had Fred Hendrickson spike posts at intervals on to the fallen tree trunks and a rope stretched across gave the most squeamish a feeling of security. Our cook took full advantage of this makeshift bridge by undressing and sitting on it on real hot days and fishing from it. A cool breeze off the water made it doubly pleasant.

Reconnoitring the Sheep Creek Trail

I decided to explore the adjacent country in hopes of locating a suitable trail to Sheep Creek, so laid out enough work down the Smoky River to keep the crew busy for a few days. This was temporary work, as we would cut the trail to Grande Cache on our way home at the end of the season. I took enough food to last me for two days, a blanket and an axe, and started off.

The first three miles were nice going, mostly along grassy side hills with game trails here and there and I was beginning to wonder why big game hunters had not gone this way instead of choosing the nightmare trail they were now using. I soon found the reason for this as I gained altitude. The higher I climbed, the more abrupt and numerous were the canyons interspersed along the river. Some had to be crossed with great caution, and at this point I should have called it a day and returned to camp as I was fully convinced that a trail along this route was out of the question. But I reasoned that this broken country might improve soon and if so, the comparatively easy going of the first part would compensate for the hard work required here. I found out how the Muddywater got its name, as one turbulent creek that I had great difficulty crossing – I could see where it had its source at the foot of a small glacier – was polluting the perfectly clear river with a stream of chocolate-coloured water. The terrain did not improve, in fact it got worse, but by now I had no thought of turning back.

Jack Glen's trophy sheep head.

JACK GLEN PHOTOS, ALBERTA FOREST HISTORY PHOTOGRAPHIC COLLECTION

I remembered having noticed a trapper's cabin close to where we crossed the river when I was with Ernie Harrison, and thought even if I arrived late it would be better than sleeping outside. A single blanket does not keep a person too warm at high altitudes and besides I might as well see some new country. Soon I came to an area that had been burned over. What a mess it was. Evidently the fire had not been hot, as the trees had been scorched only enough to kill them and had all blown down. They were criss-crossed and piled to a depth of ten feet in places, but this had not discouraged the mountain goats. They had crossed this mess both spring and fall as could be plainly seen by their foot prints and bunches of hair torn from their coats in passage.

To avoid this, I started to climb the mountain, planning to get above the blowdown and travel along the face of the hill until I got clear of it before dropping back to the river. It was a tough climb but I figured I was well repaid by the view from the top and the sight of a bunch of rams quite unconcerned by my presence. I studied them for some time and saw what I figured could be a record head, but he would be perfectly safe as long as he and his mates stayed there. No hunter would dream of taking a head or meat out of a country like that, and I was convinced that these animals had never seen a human before.

I had my lunch and rested for an hour or so, scanning the mountain peaks far and near to try to pick out individual named peaks. The only two I was reasonably sure of were Mount Bess and Mount Sir Alexander. Smoky Mountain hid my view to the southeast and that probably was the reason that I did not see Mount Robson. From my vantage point I picked out what turned out to be a pretty fair route down to the river, but I decided to keep it a secret as I made up my mind that I would pay a return visit and pick off one of those magnificent rams. Eventually I did that, and I still have the trophy.

The trapper's cabin was a welcome sight when I got there as I was utterly played out, so much so that even a promising-looking trout stream failed to tempt me. I always carried fishing tackle with me except for a fishing rod, using a willow pole in its place.

On the way back to camp next day, I decided to try for a better location for the trail than by going away up to that goat camp and floundering through that awful muskeg. After several disappointments, I finally found a more promising location and blazed a preliminary trail. This had all taken time, another day was almost spent, and I was hungry as I had finished all of my food at noon that day and still had a long way to go. When I emerged to a high point on a rock bluff and looked down into the Smoky Valley, somehow I did not feel tired any more. It certainly was a peaceful scene. I could see our camp, the horses grazing in the meadow away down the river, and two buck deer picking their way with dainty steps to a salt lick, no doubt, in the cool of the evening.

The Geological Party and the Strange Fossil Discovery

During my absence a geological party had camped near our site. They were prospecting for oil and searching for fossils, which I understand indicate the presence of oil. We were glad of their company and in the evenings we went over to their camp and swapped yarns over a huge campfire. They climbed Smoky Mountain from every direction and found several fossils.

If they had arrived two weeks later I could have shown them a whole mountainside of fossils. As I noted earlier, I had determined to avoid climbing up to the "Goat Camp," so we branched our trail off about a mile below it and on this stretch of trail to cross a steep ravine, we had to grade a series of switchbacks. Our supply of dynamite was limited, most of the grading had to be done by hand, and on this stretch we found the fossils. This was mostly shale limestone, and the fossils mostly of the leaf variety, but one specimen was different. It was about the size of a brick and had a perfectly round hole through it just as though it had been drilled. At the bottom of the ravine the banks of a small creek were entirely black shale and imbedded in them we found numerous perfectly round stones ranging in size from a child's marble to a baseball. They were quite hard and required a good sharp blow with a hammer to split them open. I have been told they bore a small percentage of silver.

A Dangerous Trail

This was an exceedingly dangerous piece of trail, being only wide enough to allow the passage of horses in single file. Several years later, an outfitter was taking a big game hunting party over to Sheep Creek. He had told the hunters about this piece of trail, so one of them rode ahead with his movie camera to film the descent. When almost opposite the cameraman, one of the horses bit the one ahead of him on the rump causing him to jump and lose his footing. The hunter photographed a scene that was not exactly in the script – that of a horse plunging to its death a thousand feet below. I understand that nothing was worth salvaging.

Although we abandoned the old trail, we spent a few days making it passable to a point where another old trail branched to Boulder Creek, which flowed into the Smoky. I had intended to do some work on the Boulder Creek Trail after we finished the Sheep Creek Trail, but we did not have time.

Sheep Creek looking towards mountains, c. 2005.

SUSAN FEDDEMA-LEONARD AND THE WILLMORE WILDERNESS FOUNDATION

Climbing out of the Smoky Valley took some time as a fire many years before had killed the original stand of timber that was practically all on the ground, and a healthy stand of second growth consisting of jack pine, spruce, willow and alder had sprung up. All this made trail blazing difficult and, as there was only one place where it was possible to get on the plateau without running into dangerous rock, I had to be exacting. We ignored the old trail since all the twists and turns it made added greatly to the distance, which I wanted to avoid. I finally located a large standing tree not far from the top and used it as a landmark, enabling us to have a pretty direct route. From there to the Dry Canyon, except for the switchbacks, the going was not too bad. This part of the trail (the Dry Canyon) always fascinated me as I often thought what a wonderful setting for a western movie it would make.

Sheep Creek Pass

The Sheep Creek Pass as far as the provincial boundary was fairly open country and a very easy climb. Our greatest difficulty was in finding solid footing, as the swamps were plentiful. It was surely a hunter's paradise. After you leave Famm Creek, which flows into Sheep Creek just below the ford, you get out of sheep country and find goat, caribou and grizzly in great abundance. Grizzly diggings were everywhere, but strangely enough all the time we were there we never saw one. We probably made too much noise in our work.

I had the good fortune to shoot a caribou bull and brought one hindquarter down to camp. The next evening two of the boys went back to pack in the balance of the meat, but a wolverine had been there ahead of us. He had dragged the other hind quarter off to his lair and defiled the rest.

Wolverine Encounters

Harold Lake[28] told me an incident that happened on his trapline in the Berland River area. An animal that is much despised owing to its depredations is the wolverine. He resembles the bear in many respects and can be just as dangerous if cornered. The Sheep Creek country is at high altitude and is a deep snow region where snowshoes are standard equipment. Harold had been doing all right until a wolverine invaded his line. It destroyed his furs and robbed and sprung his traps until he was on the verge of giving up and pulling out.

One morning he discovered a wolverine in one of his traps. The high-altitude timber there is very scrubby and not very tall, with the lower branches interwoven and almost waterproof and you might also say snowproof. This leaves a conical snowpit at the base of every tree. It was in one of these pits that the wolverine had been trapped. Harold was overjoyed and without stopping to remove his snowshoes, he drew his belt axe and was about to administer the coup de grâce when suddenly he slid headfirst into the hole. The brute snarled and lunged at him, but he was helpless to extricate himself from the snowshoes. Fortunately for Harold, the wolverine could not quite reach him or he would have been torn to pieces. Harold rested his weight on one elbow and inched himself closer to Carcajou (the Cree name for wolverine) until he succeeded in getting in a blow on the animal's sensitive nose. After that he was quickly disposed of.

One year, early in the spring I was on my way to Smoky River when I came to a place where a wolverine had been trapped. As far as the trap chain would allow him to reach, he had chewed every tree into small fragments and had broken several teeth, so the trapper told me, in his efforts to chew the chain.

First Grizzly Kill

Labour Day was a statutory holiday, so I decided to take a grizzly hunt and headed for a spot named Donald McDonald's Camp. This was some distance from where we were camped and there the grizzlies ought to be at peace with the world as there were as yet no hunting parties in that part of the country.

As I passed through a narrow gap in the mountains, I noticed a lone billy goat perched on a ledge, evidently paying no attention to me. I proceeded with great caution for another mile or so as I did not want to come face to face with a grizzly. They have wonderful vitality and have been known to kill the hunter even after being shot through the heart.

[28] Harold Lake came to this area from Nova Scotia as a youth. He was a trapper whose line was in the Smoky River area. He also had a cabin on Sheep Creek about ten miles from Clarks Crossing. His trapping area lay next to that of Cliff Faulk. Harold had a cabin at Entrance and also worked seasonally for "forestry". (Feddema-Jones 2007, Gilliat 1998)

Forest supervisor Stan Clark's pack string fording the Smoky River, 1915.

DOMINION FORESTRY BRANCH, ALBERTA FOREST HISTORY PHOTOGRAPHIC COLLECTION

Donald McDonald

Donald McDonald was a Métis from the Red River area of Manitoba who arrived in the Grande Cache area in 1905. He married Louise (Findlay) Thappe. He was a renowned bushman, travelled extensively in the Athabasca Forest region, guided and trapped. His trapline cabin was located on the meadows of Donald Flats. Neil Gilliat wrote about forest ranger Bill Smith who said he knew the trapper Donald, after whom the flats were called, and passed on stories about him. The well-established local name for these flats was officially approved in 1947. He is credited with finding the coal deposits on the Smoky River. In 1910 he showed them to Dr. Hoppe and they struck a verbal agreement for their ownership and development. McDonald cut a winter road to them in 1912. McDonald served with the 218th Regiment from Edmonton in WWI, and upon his return in 1919 found his rights to the deposits were not recognized and they were now the Hoppe Coal Claims.

(Feddema-Jones 2007, Neil Gilliat 1998, Karamitsanis 1991, Wuorinen 1997)

I sat down and ate my lunch, and decided to call it a day. As I passed the place where I had seen the billy goat, I looked up and finding him still there, I decided to have some sport with him. I have heard it said that it is impossible to get a goat excited, so taking careful aim I tried to come as close as possible to him without hitting him. He got up leisurely and stretched himself, apparently quite unconcerned. My second shot smartened him up a bit, probably stinging him with a fragment of rock, and he started to run. With each succeeding shot he shifted gears and when he finally disappeared over the skyline he was going at a mad gallop.

Some time later I went on another hunt, hoping to get either a deer or a caribou. I was well above timberline and had decided that if I intended to make camp before dark I had better be going. I took a final look through my binoculars and spotted a fine-looking caribou bull. He was across a deep ravine and too far away to try a shot, as I hate to think of wounding an animal and not being able to get him. His actions aroused my curiosity, as he seemed to be afraid of something. He would run for a distance, then stop and look behind him and run again. I was reasonably certain that there was not another hunter within fifty miles of me, so I couldn't understand why he acted that way.

The bull stopped again and all of a sudden there was a young three-year-old grizzly behind him. It was still evening and the bull gave a snort or almost a whistle, which I could hear quite distinctly, and then left for far distant points.

Junior looked longingly after him, but he must have realized he didn't have a Chinaman's chance of overtaking him. In addition to being graceful, caribou are really fast runners.

I now turned my attention to Junior. He had commenced to dig gophers, but the ground was rocky and he soon gave up and began to wander about aimlessly. I made up my mind that I would get him even if it meant staying out all night. As I mentioned before, I was across a deep ravine from him, the distance was too great for an accurate shot, and I hated the thought of wounding him and not getting him.

I tore down the ravine, but climbed the other side very cautiously as I did not know where Junior might be or what sort of mood he was in. I peered carefully over the brink and did not see him at first. When I finally located him he was just disappearing over the horizon.

I raised my rifle and fired a parting shot at him, when to my amazement he wheeled around and came straight towards me. I figured that the closer he came the less distance I would have to pack his hide to camp. Then I got to thinking what might happen if my rifle jammed, so I started throwing lead at him, but I knew I was not hitting him. The light was not good and he was not a good target, especially when coming over rough ground.

I had emptied my rifle and was feverishly reloading when he stopped and reared up on his hind legs, took a look at me, and started off at a tangent. When he was about opposite me I let him have it on the point of the shoulder. He was by this time on the edge of the ravine at an extremely steep spot and his momentum carried him out into space, and it reminded me of the song about the daring

Grizzly bear.

ISTOCK PHOTO

Grizzly bear cubs.

WEST FRASER COLLECTION

George Hargreaves (r) and Mr. Hall on the day before George Hargreaves died, 1936.

ISHBEL (HARGREAVES) COCHRANE AND THE WILLMORE WILDERNESS FOUNDATION

young man on the flying trapeze. When he wasn't bouncing he was rolling, and of course when he finally hit the bottom he was quite dead. What a relief that was. For a time he certainly had me worried.

I had only hit once as far as I could see and wonder sometimes what the outcome would have been if I had wounded him sooner. Bears of that age can be just as aggressive or even more so than their elders.

Bears mate only every second year. The cubs have the tender care of their mother and at the end of the two years might be considered spoiled brats. When the stork arrives, poor Junior is cuffed and sent out into the cold hard world to rustle for himself. At the age of three he is thus going around with a chip on his shoulder and considers himself a tough customer, which he generally is. The fact that this one ran from me instead of attacking just shows how unpredictable they can be.

I got back to camp in pitch darkness and I might say that a bear hide with the head attached is no light burden. I had him mounted with an open mouth and I still have to smile when I think of that hunt.

George Hargreaves' Premonition

Towards the headwaters of Sheep Creek there is a small creek now called Coffin Creek. It has its source at the foot of Coffin Mountain, so named because its top resembles the top of a coffin. About (1936) George Hargreaves,[29] a well-known guide and outfitter, was preparing to take a hunting party out from Mount Robson. One of his guides was helping him and when they had everything assembled, George made a trip to their cache and returned with a short-handled shovel. He went back again and came back with a sack of nails and a hammer. These were equipment he had never taken before, so the guide asked the reason. George explained that he just thought they might come in handy.

After a week's travel they camped at Coffin Creek. Now George was always an active man and a hearty eater, so when he failed to show up for breakfast one morning the guide went to his tent to investigate. He found him dead in bed. There was a doctor in the party and he found on examination that George had died of a heart attack. He had been known to say that when he died he wanted to be buried among the mountains he loved so well, so they buried him close to Coffin Creek and the shovel, hammer and nails did come in handy for digging his grave and making his coffin.

[29] George Hargreaves and brothers Frank, Dick and Roy guided and outfitted out of Jasper. George and Roy moved to Jasper from B.C. in 1912, living in a tent. George moved to Tête Jaune and opened a tie and pole cutting camp in 1917 and Jack and Frank worked with him when they returned from service in 1918. They guided and outfitted out of the Mount Robson-Jasper area. George died in the fall of 1936. As described by Glen, he is buried at Casket Creek, a headwaters tributary of Sheep Creek. (Feddema-Jones 2007, Hart 1980)

George Hargreaves' grave marker reads "Geo F. Hargreaves died September 19, 1936".

ISHBEL (HARGREAVES) COCHRANE AND THE WILLMORE WILDERNESS FOUNDATION

Abe Reimer and pack string at oufitters camp near Casket Mountain, upper Sheep Creek c. 1940s. Wrangler and cowboy Abe Reimer and his wife Mary were long-time residents of Hinton and donated land to the town which is now Mary Reimer Park

MARK TRUXLER COLLECTION, COURTESY OF JACKIE HANINGTON

Finishing the Smoky River Trail to Grande Cache, 1927

When the Sheep Creek Trail was finally completed we moved back to our previous camp on the Smoky. Our treetop cache had not been molested, which was a relief, as we were ninety-three miles from Entrance and that is a mighty long way as far as horse travel is concerned.

The first part of the Smoky River Trail was comparatively easy going as it was mostly through open meadows and along side hills, which had to be graded by hand as a heavy rain or fall of snow made them treacherous. The worst was yet to come though, as we were now confronted by a horrible mess of fallen timber resulting from a fire that hit there a few years before. People who had travelled this trail had got around it by heading right up a mountain for about a mile and then heading diagonally back to the river again. This added extra mileage to the trail, so I decided to cut right through the downed timber.

Leaving the boys to sharpen their axes and the sawfiler to file the saws I went ahead to size up the situation. We were now at a point several hundred feet above the river and almost directly opposite where the Sulphur River joined the Smoky.

Sulphur River looking upstream from the Smoky River, c. 2005.

SUSAN FEDDEMA-LEONARD AND THE WILLMORE WILDERNESS FOUNDATION

Language Problems Lead to a Life of Hardship

Glancing down at the river bottom on the opposite side, I saw an Indian walking along what was probably a trapper's trail. He probably caught a flash from the sun on my axe, as he appeared to notice me at about the same time I saw him and ducked out of sight immediately. I did not give the matter any more thought until I saw two smoke signals coming from the direction of Grande Cache about six miles distant.

It suddenly dawned on me that this was a warning for one of the tribe to lie low. If I was right in my guess, it had to do with a certain Indian who had escaped from the Fort Saskatchewan jail. He had been sent there on a charge of being drunk and disorderly, and during his stay there he had been given a team and wagon and told to work on the prison farm. A guard was with him the first few days, but the day before his sentence was due to expire he was sent out alone. The poor fellow did not know much English and probably thought he would be confined to the Skookum House for the rest of his life, so this was too good a chance to miss. It must have been quite a journey, and nothing much is known about it except that a Mountie nearly rode a horse to death following leads given by people who had supposedly seen him.

This hunt happened many years before I saw those smoke signals and at that time there were no paved highways, just trails and wagon roads. The man spoke to no one and was never seen on a road or trail. How he lived is still a mystery and how he managed to cross large rivers like the North Saskatchewan and the Pembina is unknown, but one thing known was that a dog disappeared at the same time. It is known that he came to grief crossing the Athabasca River below Entrance. At that time there was a government ferry to serve the people living north of the river. He evidently waited until the ferryman went off duty and then attempted to operate it himself. When he cast off, the handle of the winch must have flown round and smashed his wrist. He was still about eighty or ninety miles from his objective, Grande Cache, and his wrist had no attention till he reached there. It healed in such shape that he had practically no use of that hand. It was certainly an endurance test. The distance was roughly three hundred miles from the penitentiary to Grande Cache and he must have added greatly to that by avoiding towns and villages along the way.

The Mounted Police knew that he had returned to Grande Cache, but made no effort to hunt him there, knowing full well that it would have been hopeless. The minute a Mountie showed up at Entrance or anywhere else, for that matter, the moccasin telegraph would go into action immediately and the fugitive would be alerted. Indians have been known to ride a hundred miles in one day. The Mounties were really not that worried either, as he had committed no real crime. It was many years before he plucked up courage to show his face in Entrance and only then after he was sure that no Mountie was in the neighbourhood.

Using Up the Last of the Powder

To return to our trail work, we finally hacked our way through the burn and had the satisfaction of knowing that there would be no more trouble on that stretch of trail as there would be no more timber falling on it. Next we bridged a nasty little creek. It tumbled down a deep but narrow ravine and as it was at the fringe of the burn, we had a plentiful supply of dry timber at hand. We used the last of our powder, blowing out the approaches to the bridge. The season was nearly ended, and we would have no further use for explosives.

Our bridge did not last long. The following spring a slide started high up in the mountain, came down the ravine and swept everything ahead of it into the Smoky River.

All that remained to be done was a few days' work through a patch of green timber, a round-up of our horses, which were badly scattered, and the selection of a suitable ford.

Opposite page: Prairie Creek ferry across the Athabasca River, 1919. E.B. Johnson is standing on the ferry.

GLENBOW MUSEUM AND ARCHIVES NA-471-6

A Test of Endurance

The boys could finish up without any trouble, so I decided to take a sheep hunt. I started out after lunch and after a stiff climb got almost to timberline. I camped by a nice little creek and was well sheltered in a thick clump of stunted balsam. After supper I still had an hour or so of daylight so decided to make good use of it by climbing higher and scanning the surrounding country for signs of sheep. I did not see any and I started down the mountain in a hurry as it was beginning to get dark. This particular mountain had many black shale banks along its face and as I was crossing one of them I saw three large billy goats.

They were bedding down for the night and had not as yet seen me. I crouched down and watched their antics. A dog often turns around several times before lying down and that was exactly what the goats did. They also scooped out a little hollow in the loose shale, finally getting down on their knees, as a cow does, and sticking out their necks, using the shale as a pillow.

As goat and sheep seldom use the same range, I knew it was useless to hunt further on the mountain so I followed along a flower-bedecked valley till I came within sight of the next mountain, which I hoped would be a sheep mountain. I proceeded cautiously, glassing the country ahead at short intervals as sheep are pretty wary and will run at the first sign of danger.

Sure enough, I spotted their lookout. He appeared to be a good-sized ram and he was perched on a rock ledge watching the valley below. I was reasonably sure he had not seen me, as I figured the distance to be about a mile. My next move was to get above them, and I knew I had to be careful as I could not hope to get as close as I did to the bunch I mentioned before. Those had never been hunted on account of the inaccessible country in which they roamed, but this bunch had no doubt been hunted before.

Following the course of a previous snow slide I had fair climbing for quite a distance. As I neared the summit, the going began to be dangerous and did not improve as I climbed higher. Most mountains, although they appear to come to a sharp point when viewed from below, generally are fairly flat on top, but this was an exception. I was forced to crawl on my hands and knees in places as the snow and ice on the side away from the sun was frozen hard and extremely slippery. To make matters worse, a cold wind was blowing on top and as I had been sweating making the climb, I now had to proceed so slowly that I was soon chilled to the bone.

I had thought at one time of giving up and returning to the valley, but as I looked back over the way I had come it looked hopeless so I kept on. My teeth were chattering and I was afraid that if I ever let go my rifle I would never find it again, or it would be smashed to pieces in the fall. The only course left open to me was to keep on and hope for the best.

After awhile I saw what I thought was a chance to work my way down and I never hesitated. Ordinarily I would never have taken that chance, but by this time I was desperate. What a relief it was to get clear of that freezing wind and although I still had to descend slowly, I could at least think straight.

While I was up on top I did not care if I ever saw a sheep again, but as I descended and the going improved, my interest returned and I wondered if I had come far enough to get at them from above. When I first spotted the bunch they were feeding on a grassy slope just about timberline. I noticed that it was an ideal place to stalk them, as there were numerous clumps of juniper and balsam here and there.

I could not see a sign of them. Lady Luck had played a dirty trick on me and I vowed right there and then that would be the last sheep hunt for me, especially given what I had endured. Needless to say I had many more successful hunts, but none of them as dangerous as that one. I sat down and rested for a time and almost fell asleep, when suddenly I detected a movement slightly above me. Focussing my binoculars on the spot, I picked out a number of ewes, but no ram. I had overshot my mark and now had to be very cautious in getting close to them in case the ram was in the vicinity. Taking advantage of the clumps of brush, which I used as a screen, I crawled from one to another on my stomach.

Finally I located the ram and he had his little harem of ewes all to himself. I was really not in the best of condition for shooting after what I had gone through but I took a chance anyway. It was a difficult shot as he was lying down with his back to me and I did not want to spoil the meat by hitting him in the belly or hindquarters. This left no other course than to shoot him somewhere in the front quarters. While I shot he sprang up and disappeared from the ledge and I thought I had missed him. The ewes were all heading for the rocks above and I thought it was strange that the ram was not doing likewise. Then I saw him. He was below me and running with a pronounced limp. My next move was to kill him and end his suffering, but although wounded he was still making very good time and offered a difficult target. All I could do was to take snap shots at him as he appeared between the clumps of juniper. Finally he emerged no longer. It is both foolhardy and dangerous to approach wounded animals when they have fallen, for two reasons. They may struggle to their feet upon your approach and try to get away, which is only causing them unnecessary suffering, or they may, especially bears, show fight. It is much better to allow a little time to elapse and they will soon stiffen enough that it will be impossible for them to get on their feet.

This was what I did, although I was anxious to have a look at the animal. He was dead when I reached him with the final shot piercing his heart. My first shot had grazed his front leg and gone diagonally through his brisket. I considered myself lucky in not having spoiled too much meat and even having killed him at all as he could easily have escaped.

He was a beautiful animal and although not a record head, it was worth mounting. I skinned out the head and cape but as I could not pack both the meat and the head out, I took it down to the timber and hung it in a tree where I could pick it up later. The cape would have to be fleshed, salted and stretched. My only regret was that I did not have a fry pan along to cook some of the meat, which would have made a pleasant change from sandwiches. I did not realize just how tired I was until I lay down by the campfire after supper and I went to sleep right away. Some time later I woke feeling chilly, as the fire had died down. I replenished it and taking my blanket and using my coat as a pillow, I laid down on the springy bed of boughs that I had cut the night before and was soon asleep. Some time during the night I woke again to find it raining heavily. Having no canvas, my blanket would soon be soaked through and through. Taking the cape, I spread it over me. Although it smelled something like an abattoir, it was warm and I was soon back to sleep.

When I awoke at daylight I was surprised to find that the rain had turned to snow while I slept. The wood was wet and what a time I had to boil a pot of coffee. I finally made it and the fire felt so good that I hated to leave it, but as conditions would not improve and as all my food was finished, I packed up and left.

Getting down off the mountain with a load like this was quite an undertaking. On nearly all the steep downgrades that I tried to negotiate with the fresh, wet snow cover, I ended up by tobogganing down them on the seat of my pants. The going improved as I neared the Smoky River, for it was raining there and by the time that I reached our camp I was as wet as though I had fallen in the river. The boys had been worrying about me when they saw the snow on the mountains, but they did not have the faintest idea where to look for me if I had not shown up when I did.

Whistling Johnnie to the Rescue

Everything was now in readiness to start for Entrance, but as the saying goes "Man proposes and God disposes," and so it was with us. The rain, instead of letting up, assumed the proportions of a cloudburst and moving was out of the question for the day. I walked down to the spot that I had selected for a ford and found that the river had overflown its banks, so fording was out of the question. Now I was really worried. How was I going to get the crew, their belongings and the camp equipment safely across a mighty river in flood?

I remembered that the Indians had an old dugout canoe somewhere near here, but a search on our side of the river failed to uncover it. Harold Lake volunteered to cross the river and fetch it and he explained how he intended to do it. If we built a raft he would cross on that. I did not like the idea but it was a case of Hobson's Choice, as we only had a limited time to get back to Entrance.

We constructed a small sturdy raft and fashioned a paddle to go with it. Harold never hesitated a minute, but grabbed the paddle, stepped aboard and shoved off. The small craft bobbed like a cork when he steered the raft into midstream and was hit by the full force of the current. Naturally, he drifted a long way downstream with the whole crew running along the bank trying to keep abreast of him, but I don't know what we could have done if anything had gone wrong.

At last he managed to steer the raft out of the current, and by paddling furiously reached the other side. Just before he hit the bank, he made a mighty jump

Crossing rivers like the Smoky on improvised watercraft was a tricky pastime. Here, Roy Hargreaves, Curly Phillips and two hunters are crossing the Smoky River on a hand-hewn raft. c. 1920s.

ISHBEL (HARGREAVES) COCHRANE AND THE WILLMORE WILDERNESS FOUNDATION

The Fay Expedition used this dugout canoe to cross the Smoky River in 1914.

JASPER YELLOWHEAD MUSEUM AND ARCHIVES JYMA 84.87.74

and landed safely. The raft continued on its merry way gathering speed as the current hit, but it had served its purpose.

Harold found the canoe and was soon back with us, but our problems were not yet solved, as two men were required to man the canoe when loaded. Fred Hendrickson was a good boatman and I could handle a boat, but a canoe called for a different technique. Harold said that he was willing to tackle it alone by loading light, but I felt that this was imposing too much on good nature as the camp equipment and tools alone weighed over a ton. The saddles and blankets would not be far short of fifteen hundred pounds, and two men at a time was all that it would be safe to cross with.

It was still raining, but not so heavily, and I figured that by next morning the river would have risen possibly five feet more. We moved everything that would not be damaged by rain, such as the tools and saddles, down to the crossing and covered everything with canvas. This would save time in the morning as I decided that we would move rain or shine. Provided that everything went all right, our next stop would be at the Grande Cache Cabin where we would not have the bother of pitching tents. From the river to that spot was six miles, but if the distance was short I was confident that the day would be long. Little did I know just how long it would turn out to be.

Next morning the rain had stopped, so right after breakfast I dispatched the horse wrangler and one other member of the crew to round up the horses. I wanted to leave the tents up as long as possible in order to dry them out some. Ordinarily two tents made a load for a horse but in their present sodden condition one would make a load by itself. I was packing stuff in boxes when I heard someone along the trail whistling merrily. I thought this was a bit strange because most of the crew were anything but happy at the prospect of crossing the river.

As the whistler appeared in sight I recognized our old friend Johnnie Mayhoe. He was wearing his usual big grin and extended his hand greeting me in Chinook with "Klihyu Tillicum," which means in English "How do you do, friend?" I asked him what he was doing, and he said he was out looking for his horses. Knowing that an Indian is always hungry I invited him into the cook tent and told the cook to give him something to eat. I asked him if he wanted a job and explained that I needed a canoeman for the day. He was quite willing, and now one of my problems was solved, as he was truly heaven sent.

Around noon, the boys showed up with some of the horses and as soon as they had lunch they went back to find the remainder. We then broke camp and Harold and Johnnie started on the gigantic task of ferrying the stuff across. They made a wonderful team and as they gradually got things moved my spirits began to rise.

We were soon at a standstill as the boys had not yet returned with the horses nor did they appear until it was nearly dark. Getting the idea to the horses that we wanted them to swim surely tried our patience, but finally one took the plunge and the rest followed suit. Two of the boys caught and tied them up on the far side and now all that remained to be done was to get the rest of the men across.

I was a bit worried about the cook as he was the excitable type and I was afraid he would panic. Harold spoke to him sharply and vowed that he would hit him over the head with his paddle if he made one false move. He also told him not to look at the water. Being a devout Catholic, the cook crossed himself and counted his beads when he got into the canoe. His face was white as chalk and I really felt sorry for him. I decided to make the crossing with him and kept talking to him to try to cheer him up. I'm afraid that he paid no attention to what I was saying but he followed Harold's instruction by keeping his head bowed and

Horses coming out of the upper Smoky River near Glacier Creek, the river alraady deep enough to require swimming through the main channel. Forester C.H. Morse inspecting the Athabasca Forest, c. 1912.

DOMINION FORESTRY BRANCH, ALBERTA FOREST HISTORY PHOTOGRAPHIC COLLECTION

his hands over his eyes. By the time we reached the other side he was doubled up just like a snail in its shell.

I paid Johnnie off with whatever food we could spare and I will never forget the sight of him as he stepped into the canoe and started back across the river, singing at the top of his voice, "Oh it ain't gonna rain no mo,' no mo,'" which I thought was most appropriate. I for one had had all the rain I wanted in the last few days.

The cook was sick when he stepped out of the canoe, brought about no doubt by his fear of the river, so I left him alone with his troubles. But as soon as his party was over I showed him a huge pile of driftwood that had collected on a sandbar close by and told him to set it on fire. He gave me a quizzical look. "My God, you surely aren't going to camp here, are you?" he asked. "No, Fred," I said, "we need light to see to pack by." It was by this time quite dark.

The Diamond Hitch

The diamond hitch is still the preferred method for securing camping equipment and other supplies to a horse's packsaddle. Purportedly invented by a Mexican muleteer, it is a challenging hitch to tie correctly and is often used as a test of a wrangler's skill.

BOB STEVENSON

Horse Problems on a Dark Trail

We finally got everything loaded and I mounted my horse and took the lead, hoping that our troubles were over.

The trail was very steep, but everything appeared to be going all right when there was a sudden commotion behind me with cries of "Stop him! Stop him!"

I swung my saddle horse broadside across the narrow trail just in time to stop a wild-eyed bronco that we called Moose. I saw by the light of a match that he had lost part of his pack. What had happened was that the drill steel that we used in blasting had slipped back and prodded him on the rump. He registered his disfavour by bucking and trying to get rid of it and the rest of his load. When we found everything that he had unloaded and repacked him, I led him behind my saddle horse for the rest of the way.

When we packed back at the river we could not get our diamond hitches as tight as we would like on account of the ropes being wet, and this caused the pack on a spunky little horse we called Dynamite to turn under his belly. He likewise commenced bucking and kicking and finally got our dishpan loose. It was still tied to the end of a rope and was dragging just in good range of his heels.

We were now close to our cabin and also the Indian rancheree. Bedlam broke loose, what with Dynamite kicking on his dishpan and the Indian malamutes making the night hideous with their baying. It must have struck terror in the breasts of the Indians lying in their lodges as they are very superstitious, and it was also around midnight, the witching hour.

We finally cornered Dynamite and got him unpacked as well as the others and then, as everyone was ravenously hungry, we helped the cook get a meal ready. There certainly had never been a dull moment during the last fifteen or sixteen hours. That day seemed to be the end of our bad luck and the rest of the journey to Entrance was uneventful. This was one time that I hated to see the boys board the train for their respective homes because there would be no more trail crews.

I had been granted permission to build a cabin at Mile 58, so we did a bit of work on that, such as clearing off buck brush and cribbing a spring for our water supply. We made a very comfortable camp there as we built a log wall like Sam and I did at Eagles Nest and stretched our tent over that.

If road access was reasonably close, rangers would sometimes transport lumber for cabin construction using horses, such as this crew hauling lumber to build the Coliseum Lookout near Nordegg, 1927.

DOMINION FORESTRY BRANCH, ALBERTA FOREST HISTORY PHOTOGRAPHIC COLLECTION

CHAPTER SIX

HARD TIMES COMING, 1928–1929

Finishing the Eagles Nest Cabin

When it was time to go out on my district in the spring of 1928, I had as my assistant a Scotsman by the name of Bill Smith, and after a trip around the district to clear the windfall out of the trail, we started work on the Eagles Nest Cabin.

Nothing had been done towards finishing it since Sam Munson and I stopped work on it. The roof claimed our attention first, as a sizeable portion of rubberoid had blown off. We brought out the two Forestry horses loaded with metal shingles and nailed them on top of the rubberoid. This made a grand roof as the rubberoid under the shingles deadened the noise of heavy rain or hail, and was also warm. Next on the program was the floor. Bill was somewhat sceptical when I explained to him that we would split the logs, notch the round side down and then smooth the surface with an adze. He said that he had had a hard time splitting wood for the stove in short lengths and could not see how we could split logs the length of the cabin.

It was not an easy job, but we did it. I did all the adzing as Bill was afraid of the adze, and in return he did all the cooking. Another trip to Entrance was necessary to pack out windows and lime for plaster.

We spent a Sunday at Rock Lake where Bill caught a lovely bunch of lake trout. We still had only a raft there, but it was better than nothing. After we finished the cabin, which turned out to be really cosy, we travelled and explored a great part of the district.

I had been granted permission to build a cabin at Mile 58,[30] so we did a bit of work on that, such as clearing off buck brush and cribbing a spring for our water supply. We made a very comfortable camp there as we built a log wall like Sam and I did at Eagles Nest and stretched our tent over that.

Animal Games

Bill Smith[31] loved animals. He had two horses and he thought the world of them. One was a pretty fair gelding and the other a large ungainly buckskin mare. Bill would sooner starve himself than see his horses or his husky dog go hungry. One Sunday morning about six inches of snow had fallen during the night. I could see that Bill was worried when he came back to the tent and asked me to come outside and see what was going on with the horses.

We walked out to where we could see them grazing in the bunch grass meadow, and Bill drew my attention to the fact that his buckskin mare, Annie, was being chased away by a black gelding of mine. As soon as she had cleared away a patch of snow by pawing, my horse would put his ears back and make a rush at her. She was of a timid nature and would trot off to a fresh spot, where the bullying was repeated. I agreed that it was an ungentlemanly way to act, all right, but she was big enough to look after herself and I could not see that anything could be done about it. When we returned to the tent Bill sat on his bed smoking furiously for a few minutes, then he picked up his fishing gear and went

[30] Also called Summit Cabin since it was located near the head of Rock Creek on a low pass to the Sulphur River drainage, which is also the divide between the Athabasca and Peace river systems.

[31] Bill Smith was a forest ranger on the Athabasca Forest for 29 years from 1928 to 1954. He worked seasonally to 1942, which enabled him to trap on his own trapline during the winter. He was a long-time resident of Entrance.

Ranger Bill Smith haying at Berland River Flats. Rangers were allowed days off to put up hay for their horses, which were required for travel and patrols.

ALBERTA FOREST HISTORY PHOTOGRAPHIC COLLECTION

out without saying a word. We had been told by a guide that there was a little lake about a mile from our camp which had fish in it. I had taken a look at it but didn't think it looked very promising. I figured that was where Bill and his husky dog had gone, so I gave the matter no further thought.

It was getting dark and I was cooking supper when the strains of Highland Laddie reached my ear, and in a few minutes Bill appeared dragging something behind him. As he got closer I saw that he had a lovely string of fish and was now in excellent humour. He told me how he had caught each individual fish. I estimated their weight at about three pounds apiece and the funny part was that they did not seem to vary more than an ounce or two. He seemed to have forgotten the grazing incident, and that pleased me considerably as it was too trifling to make an issue of.

The Warden and the Unwanted Visitor

Bill Smith had worked for a few years in Jasper National Park and he told me several incidents about the animals of the wild. A park warden was lying in his bunk reading on a hot, sultry evening with the cabin door open. He heard a sound and looking up, saw a large bear entering. The bear touched the door with his hip and it swung shut. This bear, seeing he was intruding, commenced looking for an exit. A bear invariably never leaves a tent or building the same way he enters. The window was much too small to accommodate his huge bulk after he had smashed it, so he became quite perturbed and showed signs of rage. The warden opened up with his heavy artillery, knocking the bear down, but not out.

Practically every article in the cabin was smashed, including the bunk, which up till then was the warden's last line of defence, so now he had to depend on footwork to avoid Bruin's clutches. Eventually he managed to reach the door and got out.

It was his turn now to be enraged, so he shot the bear through the broken window until his ammunition was all gone. By this time the bear was quite dead but nevertheless he went back into the cabin and pounded Bruin with his rifle till he broke the stock. What had promised to be a nice quiet evening had ended in chaos.

Other wardens were not so fortunate. One poor fellow was killed and another had his leg broken when his horse was scared by a bear and bucked him off. He managed to crawl back to his cabin in great agony and telephone for help. It was presumed that the man who was killed had got between a mother grizzly and her cubs.

A Tough Way to Saw Lumber

Bill and I spent the balance of that season whipsawing lumber so that it would be seasoned and ready for use the following year. In this work a heavy saw resembling an ice saw is used, and a solid frame is built at a convenient height for one man to work under. A log is rolled up on skids and dogged on this frame and is squared by sawing slabs off the four sides. Lining is next, and this depends on what type of lumber is to be sawn.[32] The lines must be closely followed or poor lumber will be the result. It is hard work; the man on top does the lion's share and his partner below gets sawdust down his neck and into his eyes.

[32] To "line" a log, a chalk line was snapped along the length of the log to show where the saw should run along the edge of the next board.

The snow disappeared and was followed by a wonderful Indian summer, bringing out a rash of hunting parties. Bill was sure that we would have forest fires with so many people in the bush, but I wasn't badly worried as the parties were in the charge of competent guides. One of our posters read, "If you fire the forest, you fire yourself." As guiding was their livelihood, the guides were careful. It was the local hunters without guides who gave us the most trouble, and not only the rangers, for the game wardens had their work cut out enforcing the Game Act.

After a patrol of the district we headed for Entrance with Bill going to his trapline on the Little Berland River and Donald Flats.

While I had had the trail crew out, Louie Holm had built a cabin at Big Grave Flats. Although the cabin was fairly warm it was still a long way from being finished, but a few years later my partner and I completed the job. I don't know why Louie made such a poor job of the floor. He had just roughly hewed the top of the

Whip Sawing Lumber on Lake Linderman, 1898. This system was widely used where a lumber supply was needed and no sawmills or transportation was available. While the man on top worked harder, the man below was considered to have the worst job, "eating" sawdust with every saw pull. Jack Glen made lumber for the Mile 58 cabin this way.

GLENBOW MUSEUM AND ARCHIVES NA-891-2

floor logs with an ordinary axe and this left the floor very uneven. If you dropped any cutlery it generally disappeared through a crack. I was going to say "cutlery or money," but the average forest ranger at that time was not overburdened with wealth, so our financial losses were not heavy.

To this cabin came a party of two men. One was a guide and the other a geologist. This guide was quite a character. He trapped in winter and lived by his wits in the summer and fall. On this occasion he had lured the geologist out to the Sulphur River country on the pretext of showing him a fossilized lizard of prehistoric vintage. It may have existed, but I have my doubts – although I have seen dinosaur bones found in the Entrance district. Anyway, their quest was unsuccessful, as the guide could not find the spot where it was supposed to be. The geologist returned a sadder and a wiser man, and probably was quite saddle sore as well, as I understood it was a hurried trip. The cabin floor was very uneven and the guide who no doubt was tired made his bed on the floor and in no time was asleep and snoring. The geologist and the game warden occupied the two bunks in the cabin, but sleep was out of the question for them as, in the words of the game warden, the rafters shook when "that character started to snore." When morning came, he shook himself free from his blankets, yawned, and remarked that he had had a wonderful sleep. The game warden sarcastically replied, "I'm glad to hear that, because you're the only one who slept."

Several stories have been told about the exploits of that particular guide, who could be termed in western slang, "a tough hombre." One spring while out trapping he had the misfortune to have a wolverine rob his camp and was forced to return to the settlement. All he managed to salvage was some musty flour with

The cabin at Big Grave Flats, built by Louis Holm and finished by Jack Glen, was still standing and functional in the 1950s.

ALBERTA FOREST HISTORY PHOTOGRAPHIC COLLECTION

which he made hotcakes. He ate his fill and had one cake left over, and he boasted that he had made two hundred miles on a cold hotcake.

On another occasion he was as usual out of grub. Passing a creek, he disturbed a duck sitting on a nest of eggs, but upon breaking one of them, he found it was well incubated and had reached the stage where the duckling would soon be making its debut. It is not known whether or not he ate the eggs.

When Bill Smith and I returned to Entrance that fall, I went to Edmonton for a few days to have some dental work done, and on my return the teamster and I carried on with the road building.

The Jasper Highway was now completed almost to the Jasper Park gate, but would not be gravelled for at least two more years. We could now get out from Entrance to the highway, but it was only a fair-weather road.

The Province of Alberta was now beginning to feel the pinch, as we were approaching what was termed "the hungry thirties." To curtail expenses, the government paid off the game wardens and wished the job on the shoulders of the senior rangers.

Last Patrol, 1928

As it was a lovely fall I decided to risk a trip to the Big Smoky River. It was a foolhardy thing to do, but I was lucky and got away with it. The snow was not too deep in the high passes and I was travelling light and packing grain for my horses.

Travelling through Eagles Nest Pass I saw a sight that I will not easily forget. A huge band of caribou had congregated there, no doubt with the intention of wintering in the pass. Bunch grass was plentiful and the tufts showed above the snow in places.

My horses were used to deer and moose as they quite often fed with them when their hay was forked out to them, but they had never seen caribou except at a distance. They were quite scared of this bunch and my saddle mare was trembling. Most of the caribou had shed their antlers, but the bulls were distinguishable by their silvery necks. They did not seem actually scared, just sort of bewildered as they milled around a bit. None of Santa's team appeared to be there, certainly Rudolph wasn't.

Two golden eagles soared overhead, looking for weaklings. As I dropped down into Rock Creek Valley, I picked up the tracks of a very large grizzly who followed the pack trail for about four miles, but I failed to see anything of him. I took a chance and laid over Sunday at the camp that Bill and I had used previously. We had left our camp stove there, and all I had to do was to pitch the light silk tent that I was packing and I had a comfortable camp. On Sunday morning I took an axe and fishing tackle and hiked up to the little lake where Bill made his catch. I chopped a hole in the ice and waited for results. In the meantime I ran around trying to keep warm as a cold wind was blowing, something to be expected as this was November and the altitude around 4,500 feet.

Finally, I caught a trout and had no sooner baited my hook and dropped it in the water than I noticed another prospective customer rising to the bait. He was in no hurry to bite, though, and kept swimming slowly around, viewing the bait from every angle. Although I was chilled to the bone it was fascinating to watch the antics of the trout. He must have finally decided everything was OK as he bit hard, and I had him out in jig time. I was by this time so cold that I could not unhook him and as the line was iced I shoved the whole caboodle into my pack sack, picked up the first one which was now frozen stiff and lost no time getting back to camp.

A ranger's saddle and horse blanket dries on the hitching rail in front of the Eagles Nest Cabin, 1951.

ALBERTA FOREST HISTORY PHOTOGRAPHIC COLLECTION

Cliff Faulk's cabin near the confluence of Boulder Creek and Me and Charlie Creek.

DAVE SIMPSON AND THE WILLMORE WILDERNESS FOUNDATION

If this had been British Columbia, I would have said that these fish were landlocked salmon, as the flesh was pink, and I am sure that a blindfolded person would have pronounced them salmon on taste. By their marking, however, they were definitely Dolly Vardens. The creek that flowed past the camp emptied into that little lake, but the Dollies that I caught in the creek were totally different and had white flesh. It must have been extra good and plentiful feed that accounted for the difference. The largest trout that I caught in the creek did not weigh more than a pound.

I was favoured with nice weather, quite cold at night but lovely in the daytime. A trapper was occupying the cabin at Little Grave Flats where I had been so badly plagued with mice on my first visit. He was not at home, and his cabin dues were in arrears. I left a note instructing him to call our office in Entrance and settle his account. The mice were still there, and every trap he had set contained fur. They were at an advanced stage of decay, so I reset his traps and continued my journey.

I reached the Big Smoky River two days later. Luckily I found my neighbour and friend Cliff Faulk[33] at home and spent a very pleasant evening in his cabin. Cliff trapped in that district. He told me that a few days earlier when going over his line, he had hidden from a pack of wolves who were hot on the trail of a moose. If he had had his heavy rifle with him he would have shot at them in hopes of picking up some bounty money, but he only had a .22 and thought it rather dangerous as the pack might have attacked him.

Cliff always bought a number of Irish Sweepstake tickets, so I asked him what he would do if he ever got into the heavy sugar. He replied that he would never do anything but trap. He used to cook on hunting parties in the fall.

I did not think it advisable to visit Harold Lake at Sheep Creek so late in the season as I would likely run into deep snow. On my way back to Entrance I stayed over a day at Big Grave Flats and took a walk up the Rocky Pass Trail. It was a lovely day, calm and warm, and I sat down on a jumble of rocks to eat my lunch of roast sheep sandwiches. I had almost dozed off when I heard a rustle in the rocks. There, darting around from place to place was one of the largest weasels I have ever seen. He scarcely ever took his eyes off me, and I imagine he could smell the meat in the leftover sandwich. I tossed a piece in his direction, but as soon as I moved he ducked among the rocks. He reappeared in a short time, grabbed the meat and went back again.

There was some pocket money running around loose for me if I could get hold of him. Knowing that I would blow him to pieces if I made a direct hit on him with my heavy rifle, I made a snare with a piece of string, put it over the hole and placed a piece of meat close to it. Pretty soon his head appeared, but it was evident he wasn't going to be fooled that easily. He kept sniffing at the meat, but kept his head clear of the noose. As soon as I removed the snare he pounced on the meat and retreated with it to his den.

[33] Cliff Faulk was a trapper and a long-time resident of Entrance, commonly known as "Coyote Cliff" after his coyote-trapping skill in the 1940s and 1950s. Neil Gilliat noted that the rangers and trappers referred to him as "Whispering" Cliff, since he spoke softly as if he had permanent laryngitis. His trapline was on the Smoky River at the mouth of the Muddywater. Faulk Creek in that area is probably named after him. His trapline was adjacent to Harold Lake's. Faulk was also a fiddle player. C. Falk was listed as a seasonal forest ranger in 1941 and 1942 – this may well have been Cliff – his name has been spelled both ways. (Feddema-Leonard 2007, Gilliat 1998, Hart 1980)

I placed another piece of meat close to the hole and aimed my rifle at it. The weasel was becoming bolder all the time, so I did not have long to wait. The instant he grabbed the meat I fired, taking care not to hit him fair and square. That did it. He was forced almost into the hole by the concussion of the shot and when I reached him he was to all appearances dead, but I imagine he was only stunned so I rapped him hard on the nose to make sure he was dead.

There are two breeds of weasel. The one type is small with smooth, silky fur, and the other is large with coarse fur. Both are classed as ermines.

That night at Grave Flats Cabin I had some fun with a coyote. It was still and quite cold with a full moon making the countryside quite bright. I heard the coyote yelping at quite a distance, so I went outside and imitated it as well as I could. It kept coming closer as I answered it and seemed to get more annoyed the closer it came. I had my rifle ready but even had I seen it I doubt if I could have hit it without a luminous front sight. Then I went back in the cabin as I was getting cold and left the stage to Friend Coyote.

It snowed the night that I stayed at the summit camp and I became worried that I might be stormbound. The going was heavy, and when I reached Eagles Nest Pass I thought I would make it easier for my saddle mare by getting off and walking. Anyway, I was getting cold and stiff. Hanging the bridle reins over the saddle horn I fell in behind the packhorses where I had the advantage of a broken trail to travel on. This was fine until I had warmed up and wanted to get back in the saddle again. Try as I would, I could not get past the packhorses. My saddle mare would have stopped for me, but one of the packhorses kept nipping at her to urge her on. To make matters worse I had my chaps on, which did not improve the walking any. As I became more tired, the horses increased their lead until they were out of sight, but I was not worried, as I knew they would be waiting to be unpacked when they reached the cabin.

This time I did not see anything of the herd of caribou as the fresh snow had covered their tracks. When I reached the cabin, the horses appeared to have a smirk on their faces and I had made up my mind to withhold their grain as punishment for the trick they had played on me, but then they got such a pleading look on their faces I didn't have the heart to do it – I broke down and fed them.

Arriving at Rock Lake I noticed fresh tracks and came to the conclusion that trapper Ludwig Hoff[34] must be on his line, which proved to be correct as I had a visit from him and I asked him for supper. He returned the compliment by inviting me over to his cabin for supper the following evening.

His cabin was about a quarter mile distant from our Forestry cabin and it always amused me to visit it as it was built on a sloping sidehill. He had not gone to the trouble of levelling the foundation, with the result that he appeared to be walking uphill no matter what he was doing. If he had ever fallen out of bed, he would have ended up against the stove or the door at the lower end.

Although the trip had been enjoyable, I was glad to get back to Entrance. The

Cliff Faulk's winter fur catch, 1934.

HAZEL HART COLLECTION, HINTON LIBRARY AND ARCHIVES

[34] Ludwig Hoff was a trapper and long-time resident of Entrance. He worked seasonally for "forestry" on trail and cabin building, firefighting and on lookouts. He was born in Minnesota in 1884 and came to Alberta in 1912, worked on railway construction, later operating the pump at the Hinton station. He retired in 1955 and died in 1968 at age 84. Hoff Ridge is named after him. (Hart 1980, Karamitsanis 1991)

Tom Monaghan (l) and Bert Davey, 1927.

GORDON WATT. HAZEL HART COLLECTION, HINTON LIBRARY AND ARCHIVES

[35] Tom Monaghan was a long-time resident of the Entrance area. As early as 1910 he had a shack in Prairie Creek, the base from which he hunted and trapped. DFB records show him as a seasonal forest ranger in 1915 and 1916, and again briefly in 1920 and 1921. He purchased the general store at Entrance from Roy Woodley in 1925. With pending closure of the railroad through "Old Entrance", he moved and built a new store in "New Entrance" in 1927. He ran it with his bookkeeper/accountant Bert Davey who later took over the business in 1943. Monoghan and Davey creeks are likely named after them. (Hart 1980, Karamitsanis 1991, MacGregor 1973)

thought of being snowbound was uppermost in my mind so I considered myself lucky to be safely back again.

Two of our men were leaving that fall. Louie Holm was going to farm at Freedom, Alberta, where he died a few years later. Bill Douglas also left. He promised to write, but never did, although I heard of his movements periodically. He trapped for a year or two until he tired of that, and then he went to Fort McPherson in the Northwest Territories to run a lonely trading post. It is said that he was the man who waited on Albert Johnson, the Mad Trapper, when he called at the post for mail and to buy food, ammunition and traps. Later, Johnson led the RCMP on a manhunt that had never before been equalled for cunning and endurance until he was finally shot and killed. Eventually, Bill moved to Edmonton where he married and engaged in market gardening. He died around 1959.

Changes at Entrance

Tom Monaghan's[35] new store across the river was doing a good business, considering the Depression was gradually making itself felt. The store clerk, Bert Davey, and I became fast friends and I crossed over quite regularly to play a game of pool and pass the time. In addition to handling groceries, dry goods and hardware Monaghan also stocked baled hay and grain for trappers coming in from their traplines in the winter. They had to feed their horses since no pasture was available. Each bale of hay had a small wooden tag with the weight of the bale marked on it and it was wedged under one of the wires. When the purchaser came to settle his feed account he handed the tags from the bales to the clerk and was charged so much per pound.

One day I noticed a commercial traveller evidently keenly interested in the transaction, and shortly after he turned to me with a broad grin on his face.

"Now I've seen everything," he said. "This is the place where they still use wooden money as a medium of exchange." I did not enlighten him and when I left he was still chuckling to himself.

Patrols and Cabins with Bill Smith, 1929

The rest of that winter passed quickly and in the spring Bill Smith and I were again heading for the Mountain District. We worked in both directions from Rock Lake as usual, putting the trails in shape and waiting for the feed to improve in the higher altitudes. One morning Bill was not feeling well, so I set out alone to clear the trail to Jasper Park. I did not bother with a horse as the distance was not too great and I thought a bit of exercise would do me good. I was irritable and bored, a malady known as "cabinitis" – brought about through too much confinement in a cabin. The weather had been dreadful for the last few days.

On the way back, I surprised a moose feeding on lily pads in a small swampy pond. It was in the water up to its shoulders and when it saw me it reared and plunged in an effort to gain the bank. It reached shallower water yet it could not reach the bank as it always seemed to sink into the boggy bottom. Taking pity on the poor brute, I went back a few feet and hid where I could peer out and watch it.

The minute I was out of sight, it calmed down and for the first time used a little judgement. Instead of lunging, it folded its front legs under it and in this manner was able to draw its hind legs out of the mud. Inch by inch it reached more solid footing and got all four feet on terra firma. Had it not been frightened, I am sure it would have climbed out without any trouble. I was surprised to see such a stupid-looking animal show so much common sense.

Exploring the Headwaters of the Sulphur River

Soon after this the weather improved and we moved up to Mile 58 and cut and skidded logs for the new cabin until it was time to return to Entrance. Like me, Tom Burrows wanted to explore the country at the headwaters of the Sulphur River, so as soon as we had bought our food supply the three of us – Bill Smith, Tom Burrows and I – started out. We picked a lovely camp close to the north and south forks of the Sulphur. Horse feed was plentiful, fishing was good and we could watch the antics of the mountain goats through our binoculars.

We rode up the south fork as far as the glacier at the headwaters and I kept looking for hot sulphur springs, but although springs were plentiful they were all cold. Years later another ranger and I packed out a washtub, heated the water and took baths. We didn't stay long enough to tell whether they did us any good or

Ranger Bill Smith at Rock Lake Cabin, c. 1930s.

JACK GLEN PHOTOS, ALBERTA FOREST HISTORY PHOTOGRAPHIC COLLECTION

Tom McReady's outfit on Hardscrabble Pass c. 1940s. Hardscrabble Pass appears well named, and separates Hardscrabble Creek from the upper Sulphur River.

FAY MCREADY AND THE WILLMORE WILDERNESS FOUNDATION

Opposite page: Map 5. Trails and Cabins Built by Jack Glen 1920-45.

FOOTHILLS RESEARCH INSTITUTE COLLECTION

not, but as I recall there was nothing the matter with either of us. We tried drinking the water too, but it tasted terrible. Ted, my partner at that time, remarked that all that was lacking was a swimming pool and some bathing beauties.

Our next camp was up on the north fork close to Hardscrabble Pass, and now we were in caribou country. The afternoon that we pitched camp, Tom and I took a walk some distance up the pass, where I got some snaps of caribou keeping cool on a large patch of snow in the shade of a mountain.

When we got back to camp we saw a laughable sight, for on opening the flap of the tent I saw a large porcupine sitting on Bill's pillow making a meal off a candle that he had stuck in the neck of a bottle. I tied the flap back and went to the back of the tent to chase him out, but the brute acted just like a pig when driven and managed to scatter quills all over the place before we got rid of him.

The next day we went to the head of the pass, and like Rocky Pass, I would hate to go over it when it was covered with snow. Rockslide Creek had its source here and seemed to flow fairly direct to the Big Smoky. Tom and I figured that it might be possible to put a trail down this creek to the Smoky and abandon the Kvass Creek Trail. However, after we scrambled down it for a couple of miles or so, we gave up the idea as the terrain was too rough.

Goats in the forest.

FOOTHILLS RESEARCH INSTITUTE COLLECTION

On another occasion the forest supervisor, Bill Smith, and I were riding along the trail. I was in the lead and feeling a bit drowsy when suddenly my saddle mare snorted and shied and Bill's big husky dog dashed past and took after a billy goat. The goat did not run far as the husky was fast overtaking him and he adopted the same tactics as the nanny protecting her young. He got his back against a cutbank and met the dog's attack with a thrust of horns, which drew a howl of pain. Pandemonium broke loose and I had my hands full keeping the packhorses from stampeding. The bloodthirsty husky, showing his wolf ancestry, attacked time and again, baying and howling alternately as he was wounded by the goat. Meanwhile, Bill was trying to free his rifle from a crazed horse to save his dog from being killed. The battle continued until the goat, realizing that he was badly outnumbered, turned tail and fled, hotly pursued by the husky. Bill called the dog back in vain. Away in the distance the battle was again resumed but we never knew the end of it. I'm inclined to think the goat was the winner. When we got our horses pacified we resumed our journey with Bill almost in tears over the loss of his dog. However, the husky was made of sterner stuff and limped into camp that evening, stiff and bleeding from numerous wounds.

I have never been able to figure why that goat had strayed so far from his range. It might have been understandable in mating season, but I never knew them to be at such low altitude during hot weather. All considered, we were lucky to get out of the fracas as well as we did. Horses mortally hate goats and could have stampeded, scattering our goods and chattels to the four winds of heaven.

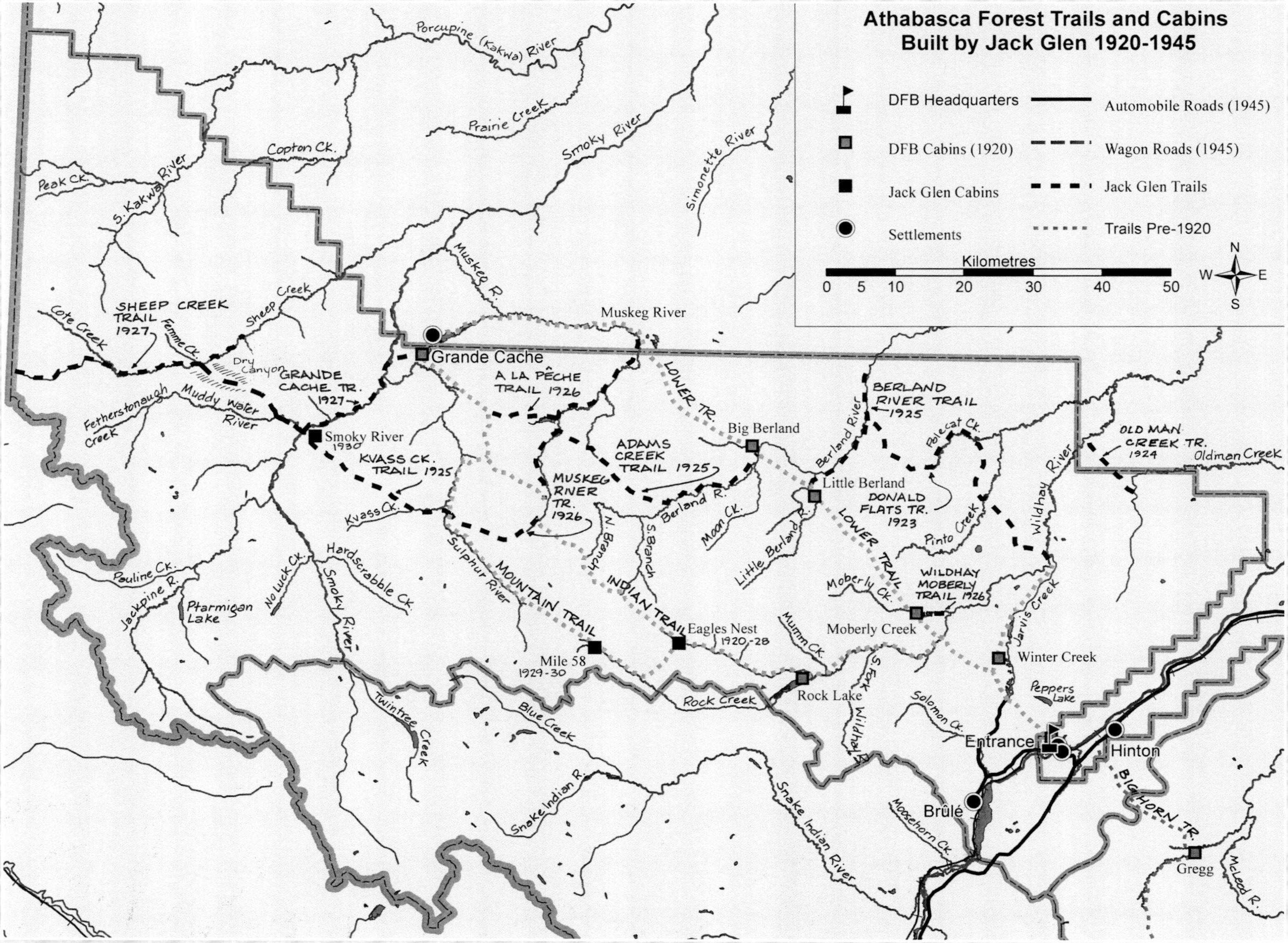

Athabasca Forest Trails and Cabins
Built by Jack Glen 1920-1945
DFB Headquarters
DFB Cabins (1920)
Jack Glen Cabins
Settlements
Automobile Roads (1945)
Wagon Roads (1945)
Jack Glen Trails
Trails Pre-1920
Kilometres
0 5 10 20 30 40 50
N
W
E
S
Porcupine (Kakwa) River
Prairie Creek
Smoky River
Simonette River
Copton Ck.
Peak Ck.
S. Kakwa River
Muskeg R.
Sheep Creek
Muskeg River
SHEEP CREEK TRAIL 1927
Cote Creek
Femme Ck.
Dry Canyon
GRANDE CACHE TR. 1927
Grande Cache
A LA PÊCHE TRAIL 1926
LOWER TR.
BERLAND RIVER TRAIL 1925
Fetherstonaugh Creek
Muddy Water River
Smoky River 1930
KVASS CK. TRAIL 1925
ADAMS CREEK TRAIL 1925
Big Berland
Berland River
Polecat Ck.
OLD MAN CREEK TR. 1924
Oldman Creek
Little Berland
MUSKEG RIVER TR. 1926
Berland R.
DONALD FLATS TR. 1923
Wildhay River
Kvass Ck.
N. Branch
S. Branch
Moon Ck.
Little Berland R.
LOWER TRAIL
Pinto Creek
Hardscrabble Ck.
Sulphur River
Pauline Ck.
No Luck Ck.
Smoky River
MOUNTAIN TRAIL
INDIAN TRAIL
WILDHAY MOBERLY TRAIL 1926
Moberly Ck.
Jackpine R.
Ptarmigan Lake
Eagles Nest 1920-28
Mumm Ck.
Moberly Creek
Jarvis Creek
Winter Creek
Mile 58 1929-30
Rock Lake
S. Fork Wildhay R.
Peppers Lake
Rock Creek
Solomon Ck.
Twintree Creek
Blue Creek
Entrance
Hinton
Snake Indian R.
Snake Indian River
Brûlé
Moosehorn Ck.
BIGHORN TR.
Gregg
McLeod R.

Cabin Construction at Mile 58

It was a lovely morning when we broke camp to return to Mile 58, but I figured a storm was not far off. Around noon we had our first warning with a vivid flash of lightning and a peal of thunder. First came the hail, then a downpour of rain which turned to snow. A cold wind sprang up and I was chilled to the bone. I dismounted and led my saddle horse for the rest of the journey, and it was lucky for all of us that I did. Tom and Bill had ridden all the way and when we got to Mile 58 they were almost frozen and could do nothing to help themselves. Seeing their plight, I dug out the dry wood and kindling that I had cached and soon had a roaring fire going. Pretty soon they got warmed up and we got unpacked and the tent pitched. With a good hot meal eaten, we all felt fine and forgot our discomforts while the storm blew itself out.

Next morning we left for Entrance where Bill and I loaded up with cement and rubberoid for the cabin roof. We were joined by Tom Coggins,[36] a forest ranger from the Brûlé District, because it takes at least three men to build a cabin.

First we made forms to pour the cement into. As it would take too much packing of cement for a solid wall, we had to be content with concrete blocks. The gravel was also a problem as we had to pack it quite a long way. Tom and Bill hewed the logs on the inside of the building, while I made the dovetail

[36] Thomas Frank Coggins was a long-time forest ranger on the Athabasca Forest from 1930 to 1950. He stayed on to live at Entrance after retirement.

Ranger crew laying the first round of logs for the construction of the Moberly Ranger Station on the Lower Trail north of Entrance, c. 1910s.

DOMINION FORESTRY BRANCH, ALBERTA FOREST HISTORY PHOTOGRAPHIC COLLECTION

corners and fitted the logs. One round a day was the best I could do and when I had eleven rounds built it was high enough with an eight-and-a-half foot wall.

For the floor, we sawed each log through the centre, notched them down and staggered them, thus making a very solid floor. We put the roof on in a hurry and sawed out the opening for the door and hung it. This was all the time we could spare on the cabin that season as Bill was to go on to the Berland District to trap in the winter months and he was anxious to get away.

Mile 58 Cabin (Summit Cabin), 1929. The cabin was located 58 miles from DFB headquarters in Entrance, near the summit of the pass between the headwaters of Rock Creek and the Sulphur River.

JACK GLEN PHOTOS, ALBERTA FOREST HISTORY PHOTOGRAPHIC COLLECTION

Fire on the Smoky River

Tom Coggins and I set out to put the trails in my district in shape. It is said that it takes travelling in the bush for an extended period to bring out the good or bad in a person, and I must say that I found Tom to be 100 percent perfect, and only hoped that he formed the same favourable opinion of me.

Crossing the Smoky

The trails took up quite a bit of our time, but we did a thorough job of cutting out windfall and repairing broken bridges as we went. I was amused at Tom when we were taking our camp equipment, food and saddlery across the Big Smoky. After I had made several trips back and forth with the boat, Tom volunteered to take the balance of the stuff across. Now I knew that this was a tricky or even dangerous crossing and that Tom had not had too much experience with a boat, but I also knew he was not the excitable type. If I objected he might get the impression that I figured he was a tenderfoot, so I gave him a few instructions and said, "Now you, Volga Boatman – take over."

The crossing was at an eddy and the idea was to follow the slack water to a point well upstream from where you wanted to land on the other side. Having gained this point you headed obliquely into the swift current with the bow of the boat pointed upstream, and by rowing like blazes you quickly reached the other side. Tom started out all right, but when he hit the current he pulled too strongly on the wrong oar and like a flash he shot back to where he started. I called him Wrong Way Corrigan, and with a grin he told me he wouldn't try it again until he received his master's ticket.

Clark's Crossing on Big Smoky River, named after outfitter and former ranger Stan Clark who built a cabin there and packed in a boat in the 1920s.

ALBERTA FOREST HISTORY PHOTOGRAPHIC COLLECTION

Douglas "Wrong Way" Corrigan

Douglas Corrigan was an American aviator. In 1938, after a transcontinental flight from Long Beach, California, to New York, he flew from New York to Ireland, although he was supposed to be returning to California. He claimed that heavy cloud cover had obscured landmarks and low light conditions had caused him to misread his compass. Corrigan, however, was a skilled aircraft mechanic and a habitual risk-taking maverick; he had made several modifications to his own plane, preparing it for transatlantic flight. Repeated attempts to get permission to make a nonstop flight from New York to Ireland had been denied in the past, and it is likely that his "navigational error" was intentional, although he never admitted it. The nickname "Wrong Way Corrigan" passed into common use and is still mentioned when someone has a reputation for taking the wrong direction. Corrigan played himself in the 1938 movie *The Flying Irishman.*

Source: U.S. Centennial of Flight Commission, 2003, online essay by David H. Onkst, www.centennialofflight.gov under "Essays/Explorers, Daredevils and Record Setters."

Stormy Weather

Tom Coggins and I went as far as Sheep Creek and on the return trip we camped in the Dry Canyon and laid over for a day. Tom went hunting and bagged a nice ram. Wild fruit was plentiful, so now we lived like kings.

The weather up to this point had been beautiful, but now it turned threatening and when we again made camp we secured everything before turning in for the night. It was just as well that we did for during the night a terrific storm sprang up and it appeared at times as if our tent might be blown away. Sleep was out of the question, as the heavens seemed to split wide open with terrific thunder and lightning and continuous pouring rain.

At the end of two days, the sun came out and the weather turned hot. It had snowed in the high altitudes and soon the slides started. I can't help but be fascinated when I watch them sweeping everything in their path down the mountainside.

We crossed at the ford on the Smoky River without any trouble where we had had to use the old dugout canoe. When we reached the Grande Cache Cabin, an Indian told us that he had seen a lot of smoke rising on the other side of the Smoky. He no doubt figured that we would give him a job firefighting as a reward for his diligence.

Most of the time I had to have Indians to fight fire when I was out in the

mountains, as one could depend on them to get to the fire in a hurry, as they travelled light and did not spare their horses. Generally they brought a teepee and, for food, a small sack of flour, tea, pemmican or dried meat, lard and a fry pan to bake their bannock on. They slept in their saddle blankets.

Porcupines were plentiful in that part of the country and the Indians were always on the lookout for them. As soon as one was sighted he was clubbed to death and taken back to camp when they went home in the evening. Their method of cooking them was simple if a bit crude. They slit up the belly, tore off the hide and quills, removed the insides, stuck them on a pointed pole and barbecued them. The length of cooking time depended greatly on the appetite, and that was generally keen. Porcupine was considered a delicacy among them, but to my untrained eye it was a difficult sight to watch them eating nearly raw meat with the blood running down their jowls.

At one fire, when we had reached the mopping up stage, one of the native men was stooped over pumping water around the roots of a tree when another lad sneaked up behind and vigorously pumped water on his rear end. During that same fire, an old hollow snag was giving us trouble and whenever the wind blew, a shower of sparks was blown across our fireguard. It looked as though cutting it down was the answer to our problem. The axes were dull and our saw had been in contact with too many rocks for the good of its teeth. Power saws were unheard of at that time. One of the lads made the suggestion that he would climb to the top of the snag and hold his hat in such a position that it would divert water from the pumper on the ground into the cavity on the snag. I didn't think it would work but I told him to try it anyway. Up he climbed and when he found a suitable limb he signalled for water. His mates on the ground did not

Opposite page: Ghost River Fire, 1915. Fighting fire in the wilderness without the benefit of roads and heavy equipment was a daunting challenge and often resolved, as Jack Glen reports, by the arrival of rain or snow to finish the job.

DOMINION FORESTRY BRANCH, ALBERTA FOREST HISTORY PHOTOGRAPHIC COLLECTION

Fire Training with backpack pumps in the Slave Lake Forest Reserve, 1925.

DOMINION FORESTRY BRANCH, ALBERTA FOREST HISTORY PHOTOGRAPHIC COLLECTION

always aim the water at his hat, as that was too good a chance to pass up when they had him at their mercy. However they made a game out of it and took turns at holding the hat. Eventually they put the fire out and although they all finished up soaking wet, they were happy.

Of course I never allowed any horseplay or pranks when a fire was actually raging. I kept track of my men, too, because when a man gets to a certain stage of fatigue he may want to sneak off somewhere and have a sleep, which might well be his last.

Unfortunately, at this time we had been instructed to hire no Indians to fight fires as they were beginning to feel the pinch and might not be above helping a fire along. I could never see it this way myself, as they must have known a fire would destroy their traplines and deprive them of their living – but orders were orders. Tom and I thought he was lying, that it was impossible for a fire to burn after all that rain, but we climbed a hill and found that it was a fire all right, and it already covered a fair-sized area.

If we could not have Indians we must have white men from Entrance, which meant that Tom must ride to the Muskeg Cabin, our nearest telephone, and summon help. When we got back to the Grande Cache Cabin, however, Bill Smith was there unpacking his horses and that saved Tom a trip to Muskeg as Bill could attend to that.

Opposite page: Fires are an agent of forest renewal, but early travellers forced to hack their way through the aftermath did not see them that way. Fay Expedition near Grande Cache, 1914.

JASPER YELLOWHEAD MUSEUM AND ARCHIVES JYMA 84.87.86

A Poor Choice for a Campsite

Next morning we retraced our steps and camped at a meadow not far from the fire. We argued about the place to pitch our tent as I pointed out that the usual campsite was too vulnerable to fire should it reach the meadow, as it had all the earmarks of doing. I thought it would be preferable to camp on the other side of Corral Creek as the fire had so far jumped the creek and although it was a horrible place to pitch a tent, it would be better than losing our outfit one hundred miles or better from Entrance. Tom did not like the idea and thought there was no immediate danger, so I reluctantly gave in.

The next morning Tom left for Grave Flats Cabin to pick up some firefighting equipment. We had a fire pump there, but not enough hose to be of any use as the fire had covered too much territory. We would have to use hand tank pumps.[37] After he left, I climbed a hill to get a view of the fire and plan the line of attack. I also found the cause of the fire. The thunderstorm had started it as lightning had struck a green jackpine and ripped a sliver, which in turn had become embedded in a dry windfall. This type of fire is known as a sleeper as it may lie dormant for as long as a week or more before coming to life. Sparks wafted by a slight breeze had found material dry enough to burn and should a strong wind spring up, this fire would soon be out of control.

The weather was again terrifically hot and dry and I was worried about our camp, as the grass on the meadow needed only a spark to become a raging

[37] Now usually called backpack pumps.

inferno. I discovered to my horror when I came down off the hill that the fire had almost reached the edge of the meadow. Racing to the camp I picked up a shovel and got back to the fire not a minute too soon as it had already reached the dry grass at one point. I tried to beat out the flames with the shovel. Whenever I thought of taking a breather I looked back and found that the fire had started up again in the grass.

I have no idea how long I kept it up but with the heat of the fire and the heat of the day I became completely exhausted and craved water. It would take too long to return to the creek, but I recalled having seen a small water hole in the timber not far away. I had the fire checked momentarily, so I made a dash for it, flopped down on my stomach and drank thirstily. As I got to my feet I blacked out and have no idea how long I lay there. When I came to, I was certain that the camp was destroyed and I marvelled that I had not been burned to death.

I had to have another drink anyway, and happening to glance at the water I discovered that it looked horrible. It was stagnant with scum floating on the surface, and it was full of wigglers. The looks of it almost turned my stomach and I headed back to the creek.

I found that a miracle had happened. The fire had burned itself out when it reached the pack trail at the edge of the timber and a change in the wind had stopped the fire from burning any more grass on the meadow, thus, no doubt, saving my life and our camp.

Unwelcome Visitors Flee the Fire

Tom Coggins returned and was surprised that the fire had gained such momentum in such a short time. We devoted our time to confining the fire to one side of the creek as the danger to our camp was now over. Visitors began to arrive – of the four-legged kind. They included a black bear that Tom promptly eliminated as he would probably have eaten our fast-diminishing food supply. Nearly every other breed of animal was represented. They appeared somewhat dazed and were seeking safety near the river. Forest fires seem to take a heavy toll of our feathered friends as we found the charred remains of many throughout the burned area. The cause of this may be that when the smoke becomes dense enough, the birds can't see in what direction to fly, and may knock themselves out against a tree or rock.

Ernie Harrison arrived alone a few days after this with a most welcome supply of food. Tom Burrows had decided that it would only be a bill of expense to the province to send out a crew at this late stage, and he asked the three of us to do the best we could. The fire jumped the creek on several occasions, but somehow or other we managed to get it stopped until a heavy rain finally licked it. We patrolled the area for a few days and put out all the spots that showed signs of life before heading for Entrance.

Harry Phillips and the Wounded Grizzly

Here I heard bad news concerning my friend Harry Phillips. About this time, the Phillips brothers were outfitting and guiding out of Jasper. Harry was guiding a wealthy American hunter. They were following the usual procedure of most hunters having shot a moose. They left the entrails and discarded meat and moved to a camp at higher altitude, hoping that the stench would attract a black or grizzly bear. At the new camp they had been successful in bagging goat and sheep and all they now lacked was bear. They returned to their former camp. Snow had fallen, which made for ideal tracking.

Harry and his hunter visited the spot where they had left the meat and found everything cleaned up, with fresh bear tracks much in evidence, and they started following them. Bruin was evidently in no hurry, as he had stopped several times and eaten huckleberries, torn up a rotten log now and again in search of ants, and seemed to have no particular destination in mind – just a "Happy Wanderer," as it were. It is wild rugged country in which they were now hunting, between Persimmon Creek and the Baptiste River,[38] and they were moving cautiously. The fresh snow, although ideal for tracking, made walking hazardous and they could not afford an accident as they were now about four days travel from Jasper and the nearest doctor.

[38] The Berland River was formerly named the Baptiste River.

They finally spotted the bear at a considerable distance from them, and as it was getting late, they decided on chancing a shot for if more snow fell, Bruin's tracks might be obliterated. The bear disappeared as soon as he was shot at, so they proceeded with as much caution as possible to the spot where they had last seen him. They found bloodstains on the snow, but Harry decided it would be lunacy to follow a wounded grizzly so late in the day and they returned to camp.

Early the next morning they returned and followed the tracks. Harry Phillips was one of the most considerate men you could meet, in addition to being a woodsman supreme. He made his hunter follow at some distance, as he knew the chances they were taking. Luckily it had not snowed in the night, and the tracking was excellent. Harry told me afterwards that there was no evidence that the bear was close. Not even a squirrel chattered to indicate any unwelcome company.

Suddenly, out of a clump of stunted balsam the bear leaped out on Harry and bore him to the ground. Harry threw up his arms to protect his head. The bear bit his hand, which caused the loss of some fingers, then he raked his body with his claws, bit at his head, but got an old felt hat, which Harry was wearing, stuck in his teeth. Trying to get rid of the hat, he reared up, snarling all the time and frothing at the mouth – gone absolutely berserk.

The hunter seized the opportunity and a well-aimed shot put the bear out of commission. As if Harry had not had troubles enough, the bear fell on him and in his death throes practically tore the clothes from his body. The hunter had great difficulty in rolling the animal off Harry, but finally succeeded in doing so.

He covered Harry with his coat and started back to camp for help, firing his rifle at intervals as a distress signal.

They got Harry back to camp and did everything to ease his suffering. He was running a temperature and they were afraid gangrene might set in, so a rider was dispatched to Jasper to have a doctor meet them on the trail.

Then the slow difficult journey began. Harry was so weak that he had to be supported in the saddle by a man on either side. They eventually got him to Jasper, where he recovered, but his hunting days were over as he was just a bundle of nerves.

Harry met his death some time later working in the woods. The following account of the accident appeared in an Edmonton newspaper under the heading, "Bachelor Killed By Falling Tree."

> Harry Phillips, 55-year-old Jasper bachelor, was accidentally killed Wednesday morning by a falling tree according to word reaching the city today. The accident occurred on the Maligne Lake Trail about three miles from Maligne Canyon. The well-known Jasper Guide and Outfitter was cutting trees for use as telephone poles, according to information received here by the RCMP.
>
> Geoffrey Wilson, who witnessed the accident, told police that Phillips was attempting to fell a tree that had become lodged between two other trees. The tree toppled and struck the Guide on the head, killing him instantly. Both men were temporarily employed by the Parks Department. The deceased is a brother of the late Curly Phillips, killed in a skiing accident at Pyramid Mountain some time ago. For years the two brothers were among the best known guides in the Canadian Rockies.

Opposite page: Harry Phillips after his encounter with a bear, c. 1929.

HAZEL HART COLLECTION, HINTON LIBRARY AND ARCHIVES

By this time there were quite a few women around Entrance, and what had been hitherto a bachelors' haven was now completely changed. At the little country store we bachelors had been content to accept anything that Bert Davey – himself a bachelor – passed over the counter, but now everything had to be choice quality or back it went.

Taffy Pull at Entrance, 1920s.
Ernie Harrison's wife Blanche at right.

ERNEST HARRISON COLLECTION, COURTESY OF LUCY MCRAE

CHAPTER SEVEN

CHANGED CIRCUMSTANCES 1930–1931

I was married in the spring of 1930. That was the first and only year that summer leave was granted to the forest rangers. The Forestry officials figured that when leaves and vegetation were lush and green there was less danger of fire. I heartily agreed with them and my wife and I spent our honeymoon on Vancouver Island and the Pacific coast. While on our honeymoon, my wife and I visited her uncle and aunt at Vernon and saw a piece of property that could be bought cheaply. I lost no time securing the title. My wife had been born and raised in Manitoba and was fascinated by the mountains. She grew to love them as much as I did.

By this time there were quite a few women around Entrance, and what had been hitherto a bachelors' haven was now completely changed. At the little country store we bachelors had been content to accept anything that Bert Davey – himself a bachelor – passed over the counter, but now everything had to be choice quality or back it went. The boys had to watch their language now, and drinking parties were less frequent. Religion, hitherto unheard of around there, began coming our way.

In the month of April 1931, our son, Jim, was born. When he was old enough, his mother would walk out with him on the trail to meet me as I returned from a trip. I would lift him into the saddle and he would ride home triumphantly. That was always a red-letter day in his life.

I remember the first minister who used to conduct services periodically at Entrance. He reminded me of Oliver Goldsmith's famous parson in *The Deserted Village.* "A man he was to all the country dear. And passing rich on forty pounds a year." This very aptly described our minister. I met him for the first time on a winter evening when he came from Edson to conduct services. The stretch of road from Entrance to the Jasper Highway was in bad shape and I met him

Oliver Goldsmith

Oliver Goldsmith (1730-1774) was an Irish writer whose most famous works included the novel *The Vicar of Wakefield* (1766), the play *She Stoops to Conquer* (1773) and the pastoral poem *The Deserted Village* (1770). He is also believed to have written the classic children's tale *The History of Little Goody Two Shoes. The Deserted Village* was a pastoral poem that lamented the depopulation of the countryside during modernization, and how it destroyed the land which former small farmers had worked hard to maintain. The vicar of Auburn is no doubt the person referred to by Mr. Glen.

Source: *The Columbia Encyclopedia*, 6th ed., 2001–2005, available online at www.bartleby.com/65/go/Goldsmit.html.

to help carry the small portable organ he used at the services. He was sensibly dressed for cold weather with a mackinaw coat and heavy pants. For footwear he had on a pair of lumberman's rubber boots, the largest of the kind I have ever seen. He appeared a trifle ungainly and except for his clergyman's collar, he could easily have been mistaken for a logger. But if his feet were big, so also was his heart, and I heard it said that he spent most of his salary on the poor and needy in his parish.

The service was held in the parlour of the Forestry building with a remarkably good attendance. His preparations were simple. First he discarded his mackinaw and replaced it with a cassock and surplice. Everything went fine until he knelt to pray, or rather when he attempted to rise up again. One knee was on the edge of the surplice, and the freshly waxed floor afforded no grip for those huge rubbers.

I felt sorry for his predicament, but I could not help seeing the funny side of it, and had the hem of the surplice ripped just a teeny-weeny bit, I'm sure I would have exploded. However, one of the congregation came to his assistance and the evening was saved.

Cabins, Messy Campers and Treed by a Bear

I had a new partner, Ted Hammer, with whom I spent many happy seasons exploring our district. He packed a small radio, which helped to while away the long evenings. We finished the Summit Cabin and I asked for and received permission to build a cabin on the Big Smoky.

Tom Burrows had resigned and our new supervisor was Fred Edgar. He was a keen sportsman and we had some very successful sheep hunts. He always hated to leave the mountains and return to his office in Coalspur. When I proposed the new cabin, he told me that money was tight and that there was no appropriation for this, but if I could build it for nothing to go ahead.

This was something of a challenge to Ted and me, so we determined to show Mr. Edgar that it could be done. Our salary could not exactly be charged against the building because if we were not doing that, we would be doing something else, so we set to and worked long hours so that none of our other work would be neglected. There was a block of fire-killed spruce not too far from the site and there were ideal building logs in it. Ted skidded them in with his saddle horse as they were well dried out. We did not whipsaw the lumber as that would have taken too long. For the floor, we used the same construction that we used at Eagles Nest – we split the logs down the centre and I adzed them smooth. The roof was of shakes nailed to poles running parallel to the building and the walls were chinked with moss. All the furniture and the door was made of slabbing logs, which were adzed and planed smooth. The two windows we packed out from Entrance, where they had been in the attic of the cache, forgotten for years. The total cost of the building amounted to only a few pennies, the cost of door hinges and nails. Mr.

Ted Hammer

Ted Hammer and his brother Harry were raised on a farm in central Alberta. Ted joined the Dominion Forestry Branch as forest ranger on the Athabasca Forest in 1930, transferring to the Alberta Forest Service that fall with the Transfer of Resources. In the fall of 1940 he moved to Grande Prairie when he was promoted to the position of Timber Inspector. He was appointed head of forest protection in 1955 and moved to Edmonton where he retired in 1964. When funding became available in 1959 to build a Forestry Training School, Ted drew on his experiences on the Athabasca Forest to urge that the school be located at Hinton and on the site of the new ranger station. Later, the quarter section for the old forestry pasture west of Hinton was traded to the Town of Hinton for additional land around the school to enable expansion in 1964. The forestry pasture became the Hinton Golf Course.

Hunter with moose head for mounting, c. 1940s.

FRED TIETGE COLLECTION, COURTESY OF STAN TIETGE

Edgar laughed when we told him that the cabin was finished and he admitted he didn't think we could do it. He promised us that we should at least have a good cook stove to go with it.

Ted and I enjoyed our latest luxury. Where we had camped formerly was becoming a regular breeding ground for mice and other vermin. They would eventually find their way into the cabin, but we figured that we could eliminate them easily enough with our wheel-of-death trap.

The mechanism of this is simple. First a large can is required. Ours was a five-gallon kerosene can. The lure is an evaporated milk can with a wire spindle through the centre, placed halfway across the top of the large can and smeared with lard or bacon grease. A board from the floor to within a short distance from the revolving can completes the trap. The mice, attracted by the grease, jump from the board onto the milk can, which in turn revolves and dumps them into the water in the bottom of the large can, where they eventually drown. Ted figured it would be more humane to have life jackets for the little fellows, but I could not see this.

We had a visit from the game guardian before we left the Smoky Cabin, so we arranged to take a grizzly hunt. Constable Archie Clark[39] acted as game guardian in addition to his regular police duties, and he jokingly said that he acted as game guardian while he was resting. We had much in common, Archie and I, and I was sorry to see him transferred after a year or two in that district.

[39] Archie Clarke was an RCMP Constable for the Entrance and Coal Branch area and was also a game guardian. He became the first policeman of record in Hinton and, with an assistant, shared an office behind the Hinton Hotel. Ranger Neil Gilliat rode game patrols with Clark in the 1940s. (Hart 1980, Gilliat 1998)

We had a little target practice with Archie's service revolver before leaving on the bear hunt. Ted was invited to take part, too, and at first he refused. "What chance have I against two old buck constables?" he asked. However, he finally gave in and after we taught him the proper way to shoot, he did very well. At first he shot like the bad men in Hollywood movies until we told him to pull up on his target instead of throwing down on it.

We had a very enjoyable trip, but did not even see a grizzly. The fishing was good, however, and that helped to compensate. We travelled as far as the interprovincial boundary and then rode a few miles into British Columbia. The weather turned cold on our way back, but since we had good eiderdown sleeping bags it wasn't too much of a hardship. We had a rousing campfire in the evenings and I would brew a pot of cocoa before we went to bed. I did the cooking and I think that was the best job, as Ted and Archie rounded up the horses in the morning, sometimes getting wet feet and clothes in the process.

Ted and I decided that we had to travel separately on the job as there was a certain amount of lawlessness going on and campfires were left burning just as soon as we turned our backs. Since we were going and coming at different times, the vandals could not keep track of us and pretty soon we had no more trouble.

On one occasion we had to double up, as we had some heavy work to do along the Sulphur River. We left Grave Flats Cabin bright and early one summer morning, as we wanted to make camp before the heat of the day. This part of the

Ranger Wally Harrison at Smoky River Cabin, 1951

NEIL GILLIAT PHOTOS, ALBERTA FOREST HISTORY PHOTOGRAPHIC COLLECTION

This unique combination chair/ stepladder/ ironing board was built by Jack Glen for the Moberly Ranger Station. Donated to the AFS Archives by the Glen Family, 2006.

BOB UDELL

trail was seldom travelled and we had gone only a few miles when we came to a spot where someone had camped.

It was a filthy mess. There were entrails and hides scattered around, and the blowflies were active. Teepee poles were lying criss-crossed on the trail, which is something that you seldom find, as the native women generally stack them neatly against a tree. To make matters worse, a campfire was still burning. The tracks were fresh so after putting out the fire we hurried after the miscreants. It was a very hot day and I hated to work horses under such conditions, but it was evident that the party was headed for Grand Cache. This was a long day's travel, but we consoled ourselves that we could rest the following day.

We reached Grande Cache just as two Indians were unpacking. I will admit I was in a vile mood, but controlled myself as well as I could when I began to question them. One was a man of about sixty and the other a young lad of sixteen or seventeen, the old man's nephew, I believe.

The old man could speak some English, so I started questioning him in a roundabout way and was getting along famously until I got down to brass tacks and accused him of leaving a burning campfire. Suddenly he had no more knowledge of English. I didn't know much Cree, so I started in on the kid who lied for all he was worth and denied everything. If he had admitted leaving the fire burning I might have been inclined to give them a reprimand and let it go at that, but when he would not tell the truth I lost my patience and told him that I would be looking for them at Headquarters at Entrance at the end of the month for further questioning. He refused at first, saying they had not intended to go to Entrance until fall. I told them they had better change their plans and be there, or I would be back after them. I also told them to go back the next day and clean up the mess they had left. Ted thought that I was being too hard on them, but I told him they had committed a serious offence and must be punished accordingly. They reported to Headquarters as instructed, and received a light sentence, but I think it served as a good lesson to the whole tribe.

Boulder Creek Fire

That same year a fire had been reported on Boulder Creek, about one day's travel up the Big Smoky from our cabin. I had gone to Entrance with the diaries and monthly reports and knew nothing of the fire until I reached there. Unfortunately, Ted didn't know that part of the country at all and didn't know about an old trail that would have taken him to the fire in a day. Bill Smith joined Ted at the Smoky. He had with him three young lads who were junior fire wardens. They commenced cutting a trail upriver, which took them three days. For horse feed all that was to be found was a marsh, and a mountain cayuse – accustomed to luscious bunch grass – would sooner starve than eat slough grass. They hobbled their horses and hoped for the best. One of Ted's horses showed his disfavour by hitting the back trail, and in trying to swim the river he drowned in a canyon.

When I arrived, I found a very much-disgruntled crew as they had put in long hours cutting the trail and got very little sleep on account of mosquitoes and sand flies. Ted had sent out for a pump, and the packer had arrived via the old trail before Ted and his crew reached the fire. He could not understand what had happened to them.

The fire had been caused by a lightning strike at a high altitude and had burned down to Boulder Creek at one place and been halted by a swamp. We had nothing to worry about there, unless a strong wind sprang up and caused it to spread. It had not so far succeeded in jumping a deep rocky canyon but could do so at any minute, so we concentrated on that. First, though, I sent the pump back with the packer, as we did not have sufficient hose to use it. Our camp was about three-quarters of a mile from the worst point of the fire, with the result that we had to carry our drinking water and water for our tea at lunchtime. A thermos flask at that time was unheard of. The weather remained hot and dry and it looked like this was going to develop into a serious situation. We had pooled our food supply, but it was rapidly diminishing. To make matters worse, some friction had sprung up between two members of the party, so you can imagine my relief when we had several good smart showers. I know I was taking a horrible chance, but sooner than see any fighting in camp I sent them all back. Ted was overdue at Entrance anyway.

No sooner had they gone than the rain showers quit and I put in some very anxious hours running from place to place where the fires had jumped. After the boys left I had trouble with my horses, too, as they seemed to miss the company of the other horses and wanted to pull out. I had to tie them up through the day and turn them loose at night. I also moved my tent to a spot on the trail where they could not pass without my knowing it.

The weather began to look threatening and at midnight there was a terrific thunderstorm, followed by a regular cloudburst. By morning the snow was coming down in large wet flakes. It was a horrible day to travel, but by this time I was on a starvation diet and was anxious to get back to the Smoky Cabin where I had plenty of food cached.

The old trail joined the Sheep Creek Trail near the Goat Camp and it snowed until I got there, making travelling very hazardous. My saddle mare slipped and fell several times, and I dreaded the possibility of a broken leg. As soon as I left the high altitude the snow turned to rain and by the time I reached the Smoky I was soaked to the skin in spite of my chaps and heavy mackinaw coat. I unpacked the horses and they lost no time swimming the river and heading for the pasture. The boys had forgotten to turn the boat over, so that it was almost half full of water. I could not have been much wetter, however, unless I had fallen in the river. With dry clothing, the stove working overtime and a good hot meal, I felt like a new man. The comfort afforded by the little cabin had repaid us for the hard work that we had put into it.

Forest rangers crossing divide above timberline at headwaters of the Snake Indian River, Rocky Mountains Forest Reserve, 1914

ALBERTA FOREST HISTORY PHOTOGRAPHIC COLLECTION

Technology Changes Improve Communications

About this time the Forestry two-way radios were issued to us and we welcomed their arrival, although it meant extra work for all concerned. I was forced to buy another horse as the radio and accessories weighed almost a hundred pounds, and my horses were already carrying their limit. My call letters were VX4E, and we had regular schedules worked out. It had often worried me before that if some accident happened to someone alone, no one would know of it until it was too late to help. Now if one missed a sked or two, the others would become suspicious and investigate.

Treed by a Grizzly

A little later that year I was making a routine patrol and had stayed at Grave Flats Cabin. I had an early breakfast as I intended to make the Smoky Cabin that day. It was a hard trip and I was anxious to catch the horses and be on my way as soon as possible. I could hear my horse bells at the upper end of a long bunch grass meadow and thought how fortunate I was to locate them so quickly.

All at once a huge grizzly came out of the bush and began to dig gophers out of their burrows. He was between me and the horses and evidently had not seen me. I am of the opinion that bears are nearsighted. I figured he would move off soon, but nothing doing. Although I was impatient to be off, I watched his antics with interest. They look so awkward as they dig, sniff and listen, and then dig some more, to be finally rewarded with a dainty morsel. They get quicker results eating ants as all they do then is tear open a rotten log inhabited by the insects, stick out their tongues and when they are covered with fighting mad ants they swallow and repeat the process.

Time was flitting, so I thought I might scare him off. I started towards him whooping and swinging my riding bridle. He looked up and saw me, reared up on his hind legs, and came for me.

I made for the nearest tree and lost no time in getting to the highest limb. He stopped a few feet from the tree and looked me over carefully with his little beady eyes, and then went back to his digging. Now a grizzly can devour the hindquarters of a moose at one sitting, so I thought, ye gods, if he has to catch the equivalent of that in gophers, I'll remain up that tree till the duration. To make matters worse it started to rain and I had not brought my coat.

I would have given anything that morning to have had my rifle and been able to wipe out that insult. Bears were bad tempered that fall. This was attributed to the fact that the wild berry crop in British Columbia had been ruined by a summer frost and hunger had driven the bears over to Alberta. I did not feel so bad when I heard later that a guide had been treed. Ted Hammer, too, was forced to make a wide detour while bringing in his horses to the Summit Cabin, as the grizzly blocked his way. Apparently my grizzly did not like eating in the rain, for soon after it started he sauntered off, much to my relief.

Tony Earnshaw, Alberta Forest Service radio technician with a new "portable radio". Crowsnest Forest, 1949

ALBERTA FOREST HISTORY PHOTOGRAPHIC COLLECTION

Youth and Inexperience Yield Tragic Results

Once I was at the Summit Cabin waiting for Edson to contact me on the radio when another ranger broke in and asked me to stay on the air as he had an important message. He gave me an account of a drowning accident that had taken place at the Big Smoky somewhere near our cabin, and asked me if I would conduct a search for the body. An ex-forest ranger by the name of Harvey[40] had been given the key to our cabins and I had met him on the trail when I was going to Entrance. Before he quit the Forest Service he had been raising horses on a ranch in southern Alberta and was heading for the Peace River country where he hoped to dispose of a bunch of them.

He had with him a young lad of eighteen or thereabouts. Not being familiar with the ford on the river, he probably had attempted to cross at the eddy, which I have formerly described. It was all right to swim horses here when they were unpacked but foolhardy to do so when they bore loads or riders. The regular ford was about three-quarters of a mile upstream.

I left Big Grave Cabin the following morning and was fortunate to find Constable Archie Clark at the river when I arrived. At the Smoky we found no clues

[40] James Shand Harvey.

Horses swimming across the Smoky River at Clark's Crossing, 1957

ALBERTA FOREST HISTORY PHOTOGRAPHIC COLLECTION

Rangers and outfitters respected their cowboy heritage. Here Les Lama demonstrates rope twirling in full western dress regalia. Entrance 1927.

FRED TIETGE COLLECTION, COURTESY OF STAN TIETGE

to follow as considerable rain had fallen since the accident. It seems that the lad had got into serious difficulty and he had got out of the saddle and attempted to swim himself. When Harvey looked back he noticed the empty saddle, but there was no sign of the rider.

We found the horse later, still saddled, and he must have been in agony as the cinches had cut into the flesh and had festered badly. The horse was pretty nervous and hard to catch. Not being familiar with the country, Harvey had been unable to report the accident until he reached Grande Prairie. If he had known it, he was only two days' travel from the Muskeg Cabin, where he could have telephoned and given us full particulars.

We assumed they had crossed at the eddy, and conducted our search accordingly. Archie rode downriver and kept a sharp lookout for any sign of the body on the sand bars or at the ends in the river, and I probed around the eddy. It would not take long for eagles and wolves to scent a body washed up on shore, so we were hopeful of success. I had seen what happened to chunks of wood and floatage when it reached the eddy. It would be spun around for days sometimes before being cast to one side, and this could well happen to a body.

I took a pike-pole and jabbed here and there. In one spot there was quite an accumulation of trash with froth and scum on the top of the water. I stuck my pole underneath this and tried to dislodge something solid. Suddenly my pole slipped and I took a swan dive into the water. Luckily for me the water only came up to my breast and it was not too swift nor too cold. I had no difficulty getting out, but my watch was ruined. Archie returned late that night and reported no luck. He very kindly lent me his watch so I could keep my radio schedules. We spent another day or two searching likely looking places, but since we had nothing definite to work with we decided to give up.

A little later that season a young lad acting as a horse wrangler on a hunting party had a freak accident. He must have been reading too many western novels for his own good. Decked out in full western regalia, he had put up a target and was shooting at it with his revolver, in the meantime practising a fast draw. He got a little too fast on one of his draws and pulled the trigger before the gun was out of the holster. The bullet entered the fleshy part of his hip and emerged in the calf of his leg. He lost considerable blood and suffered considerable pain before he had medical attention. One of our rangers was saddled with the chore of bringing him to Entrance where he would be taken to Jasper by railway speeder. We were having Indian summer at the time, hot in the daytime and freezing at night, and the wound commenced to fester. Riding caused agony during the day and at night the weight of an eiderdown was almost unbearable. He had several delirious spells also, which did not make things any easier for his ranger nurse. The only good thing as far as the patient was concerned was that he had a warm cabin to go to every night. He would have suffered more in a cold tent.

The wounded youth in Jack Glen's story may have been transported on a Forest Service speeder such as this one used by Rangers Greenwood and Rance patrolling in the Brazeau Forest, 1913

DOMINION FORESTRY BRANCH, ALBERTA FOREST HISTORY PHOTOGRAPHIC COLLECTION

The Forest Service at that time believed in fire posters, which were tacked up on trees. Ranger Bob Holgate had very little bush experience, and used to put the posters on small trees as often as on large trees. Ernie Harrison saw the funny side of this small tree posting, and said that to read one of these posters, a person had to ride several times round the tree.

Patrol on a Forestry trail stopped at a Fire Notice posted on a tree to remind travelers to be careful with fire, 1920s.

ERNEST HARRISON COLLECTION, COURTESY OF LUCY HARRISON

CHAPTER EIGHT

FIREFIGHTING AND SEASONAL WORK

Ranger Bob Holgate.

FRED TIETGE COLLECTION, COURTESY STAN TIETGE

We were now entering that deplorable period known as the "hungry thirties" and to make matters worse, the natural resources had been taken over by the province of Alberta, which at that time was in bad shape. The results were inevitable. Men were laid off left and right, myself included. I had lost two years' seniority by going farming at Calmar, otherwise I would have been kept on permanently. When the lay-off came we moved to our property in Vernon and I built a house on it, hoping to find work there in the off-season.

When I was notified to report for duty in 1933, I left my wife and son in Vernon and returned to Entrance, where I put in one of the loneliest times of my life. My wife was lonely, too, and when I returned to them Jim didn't know me, so we vowed we would never be separated again.

That season, however, was not without excitement. Spring came at least one month earlier than usual and the weather turned unmercifully hot. The Headquarters' ranger had somehow, in spite of the Depression, been granted an appropriation to build a new ranger station, as the old building had started to deteriorate and was hard to heat in winter. We were pouring footings for it when a fire was reported on the Big Smoky.

Smoky River Fire

Ted Hammer and I dropped everything, bought a grubstake and headed out. We had to pass up the Summit Cabin because there was still snow there and no horse feed. It was a long day from Eagles Nest Cabin to Grave Flats Cabin, but we had no alternative. When we arrived there late in the evening it was summer weather with the grass green and a small herd of caribou playing in the meadow. When our horses saw them, tired as they were, they got quite frolicsome and as soon as they were unpacked they had their usual roll and tore up the meadow bucking and kicking up their heels. If we had not had a fire claiming our attention we would have liked to stay and rest there for a day or two, but we had to keep moving.

When we arrived at Kvass Summit the next day at noon, at first it looked as though our trip had been in vain. We were confronted with a huge snowslide that had wiped out the trail entirely. We estimated the slide to be about thirty feet high and over a quarter mile wide. Ted looked after the horses while I walked ahead and sized up the situation. I climbed up on the slide and jumped on the snow, but hardly made a dent in it. It was packed together so tightly it was like ice. Trees, rocks and mud were all embedded in it and we figured that a lot of work would have to be done when it finally melted before the trail would be made passable. We led our horses across it with no trouble and reached the Smoky without incident.

The fire by this time had covered a large territory but was still about six miles from the cabin. Bob Holgate,[41] the ranger for the Muskeg River District, was to recruit a crew of firefighters from among the Indians at Grande Cache and fight

[41] Bob Holgate was a forest ranger on the Athabasca Forest for 15 years from 1923 to 1937.

the fire from that end, as that was where it was likely to do the most damage. They decided to hire Indians again because they were right there, whereas conscripted white men would have cost a small fortune to transport. We never found out what caused the fire, but I'm sure the Indians didn't start it because it was burning on their traplines and could quite easily have wiped out their homes. It was no doubt lightning.

Ted and I busied ourselves in cleaning and burning all inflammable material close to the cabin and whatever looked likely to burn our pasture. We backfired a stretch parallel to a small creek not far from the cabin. We felt reasonably safe, but were afraid that a spark lighting on the roof might, if not seen in time, burn the cabin down.

We decided to try to make contact with the other party if possible, so we started out on foot at daybreak one morning, as it was too risky to take horses. The weather was suffocating and we travelled in our shirtsleeves and took only a lunch with us, feeling confident that we would reach camp that evening.

Ranger Holgate had no radio because he was always close to the telephone line and didn't require one. When we reached the fire on our end, we found that it was burning fiercely in a beautiful stand of lodgepole pine, so we figured we had better get to the river as soon as possible. Before we reached it, we had to cross a stretch of bone-dry windfall piled up to six or eight feet in places. Getting over this was dangerous and I warned Ted to be careful. A broken leg in a spot like that would certainly be a catastrophe. He failed to see the danger, pointing out a narrow swamp at the edge of the windfall which he thought would stop the fire. I had my doubts, but said no more.

Smoky River five miles downstream from Sulphur River near Grande Cache.

FAY EXPEDITION 1914, JASPER YELLOWHEAD MUSEUM AND ARCHIVES JYMA 84.87.73

When we reached the river, we found to our disappointment that the fire had burned right to the water's edge so when we were not walking in the water we were crossing stretches of still burning moss and logs. The heat was terrific and had we not had water to drink I think we would have collapsed. We finally emerged on a stretch of rocky ground where the fire had been halted and sat down and ate our lunch. We tried to figure out where the camp would be, but could not see far on account of smoke.

We started off again feeling dejected and tired, and at sundown reached the Sulphur River, which was in flood so any hopes we had had of wading it were off. We had hit the river exactly opposite Grande Cache. The nearest building would be about three-quarters of a mile away, so we had hopes of attracting the attention of an Indian and having him fetch a couple of horses so we could cross over. We shouted and shouted, but never a sign of an Indian did we see, and all we could hear above the roar of the water was our echo.

It was hopeless to stay there any longer, so we turned back and travelled until it was too dark to see. We started a small fire in a ravine where there was a trickle of water and tried to sleep, but that was out of the question. It froze that night, and we were certainly cold, having no coats. Had we an axe we could have chopped wood enough to stay comfortable by the fire, but all we could do was to break off small twigs, which threw off little heat. We finally let the fire die out. I gathered all the moss that I could find and Ted did the same, but it was no substitute for blankets. I kept looking at my watch and could have sworn it had stopped or was running in reverse. The first sign of daylight was indeed a welcome sight. We started off hungry and with our teeth chattering like castanets. When we tried to speak to each other it sounded like a lot of gibberish. We soon warmed up and craved food, but that was hopeless until we could get back to the cabin. Even a slice of bully beef, which I loathed, would have tasted like T-bone steak that day.

We went through the same misery that we suffered the day before, only more so, as the fire had advanced at least two miles over the whole front. When we reached the spot where I had told Ted to be careful, it was completely burned and was easy to cross now, except for breathing hot ash.

Between two and three o'clock, we reached the cabin and I cooked what was supposed to be our breakfast, but when we sat down to enjoy it we could not swallow a bite. Our throats were raw with inhaling so much hot ash and smoke and nothing would go down. Ted remarked, "If I can't eat, I sure will sleep." It was too hot to sleep inside and I took my sleeping bag to the shady side of the cabin where I was immediately dead to the world.

Ted and I could not eat anything solid for several days. We got heartily sick of puddings, soup and mush. Although the diet was monotonous, we could yet look on the brighter side of life and poke fun at each other. Ted told me I was dying a lingering death, and I got back at him by saying there was no danger of him going the same way, as he could live for many years on his own fat. (He was not really fat.)

We did all we could to safeguard the cabin and our boathouse. If the fire had got to the boathouse, we would have covered the boat with sand and would have done the same for Stan Clark's boat if the fire had jumped the river. Ours was a log boathouse and a mighty small one at that. Ernie Harrison and an assistant ranger had built it when I was in charge of the trail crew, and whether they were in a hurry to finish or had got the wrong measurements I never found out. Luckily the door had been cut slightly out of plumb and there was just one way the boat could be got out and in. A provincial game guardian once remarked that if we ever painted the boat, we would never get it inside. However it was light enough so that one person could handle it and when I was alone, I used to get it on my back, at the right angle, and crawl in and out of the building.

The fire kept gaining ground in spite of adverse winds. During lulls we could hear the spruce and pine needles pattering on the roof, and the smoke got so thick that one got the impression that the fire was right close. We had made up our minds to catch up to our horses and turn them across the river, when finally the rain came. It simply poured for two days and snowed at higher levels.

The Smoky River and Range, (left hand side) taken from the hill up the Muddywater River.

ISHBEL (HARGREAVES) COCHRANE AND THE WILLMORE WILDERNESS FOUNDATION

GOVERNMENT OF THE PROVINCE OF ALBERTA

DEPARTMENT OF LANDS AND FORESTS

WARNING

TO SETTLERS AND OTHERS

BURN YOUR BRUSH

PILES DURING THE WINTER MONTHS

NO BURNING PERMITS

WILL BE ISSUED AFTER APRIL FIRST UNTIL CONDITIONS ARE SAFE FOR SUCH BURNING

If you wish to cultivate your land next spring the brush piles must be burned this winter

Persons responsible for the setting of fires without authority will be prosecuted to the full limit of the law

EDMONTON, ALBERTA

DIRECTOR OF FORESTRY

Sunday morning broke bright and clear. We had slept late and were just finishing breakfast when one of the Indians from Holgate's crew arrived to find out what the score was on our side of the fire. They had managed to cut the fire off at the canyon on the Sulphur River just as the rain came. We looked the situation over, and I sent back a note telling Holgate to let us have the crew, though there was no need for him to come himself. I did not think he would care to put his horses over such country that an Indian thought nothing of.

Bob Holgate had been a sailor a great part of his life, and had turned to hard rock mining before entering the Forest Service. The Forest Service at that time believed in fire posters, which were tacked up on trees. Holgate naturally had had very little bush experience, and used to put the posters on small trees as often as on large trees. Ernie Harrison saw the funny side of this small tree posting, and said that to read one of these posters, a person had to ride several times round the tree.

Although the rain had halted the fire, it was still a long way from being out. The weather turned dry and windy again and we worked long hours to keep it in bounds. New fires sprang up, but were quickly controlled as the humidity was still fairly high and the ground damp.

Opposite page: Forest Service fire poster, 1940s

ALBERTA SUSTAINABLE RESOURCE DEVELOPMENT

A Stubborn Fire on the Mountain Trail

When at last I figured it was safe, I sent the crew home, and Ted and I headed back to Kvass Summit to clean up the mess that the slide had left. At first it looked hopeless, but by cutting a new trail practically all around it we got it finished. We had not forgotten our Smoky River fire, and made several trips partway up the mountain where we could look down into the valley. There was no sign of smoke, so we headed back to Big Grave Flats Cabin.

We found the Sulphur River very high and muddy, and were disappointed as we had been looking forward to some good fishing. The Big Grave Flats Cabin was built on a little creek from which we got our water supply. It was always cold and clear and where it flowed into the river it formed a small clear pool. I was walking past this point when I happened to glance down and noticed that it was simply swarming with Dolly Varden trout. I went back to the cabin to get out my fishing tackle and break the good news to Ted.

Those fish were simply ravenous and in no time we had all we needed for at least two days. After that length of time a person generally gets a bit tired of them. That sort of fishing was certainly no sport for us, so we decided to try something else. We cleaned two of the fish and tied part of the intestines on a long string and cast into the pool. Ted remarked that every fish in the Sulphur River was there that day and it certainly seemed like it. If we had no sport before, we certainly had plenty now. We kept the bait moving swiftly from side to side with the eager beavers tearing after it. Once in awhile one of them would be rewarded by tearing a fragment off the bait, and sometimes one would snitch the whole caboodle. One

Creel and lake trout from Rock Lake, c. 1930s.

JACK GLEN PHOTO, ALBERTA FOREST HISTORY PHOTOGRAPHIC COLLECTION

big fellow swallowed the whole of the bait and several inches of string besides. He apparently had his teeth clamped tight on the string for, pull as I would, I could not manage to get free of him and had to land him to do so.

It was now time to go to Headquarters with the fire report and diaries. Ted would stay out for another week and then start in. We planned to meet at Rock Lake, but that was not to be.

The distance from Rock Lake to Entrance is about thirty-two miles, and as this is considered too far for horse travel in one day, I used to break my journey at Major Brewster's round-up cabin on Solomon Creek, if it was not occupied at the time.

On my return trip, the radio reception was poor owing to thunderstorms, so I did not pack my radio past Rock Lake. I knew nothing of another fire until I arrived at Eagles Nest Cabin. Ted had found it on the strength of a report from Fred Hendrickson at Athabasca Lookout. When I reached the summit, Ted was already there as his food supply was low and he couldn't do much about the fire while alone anyway.

This fire, evidently caused by lightning, was about two miles above the Mountain Trail at a point half way between the Little Grave and Big Grave Flats.

I pitched my tent by a nice little creek and, after breakfast the following morning, thought it would be a good idea to cut a good supply of firewood. Once firefighting starts there is little time for anything else, so I sawed several blocks off an easy splitting pine tree. It proved to be a little too easy to split, for when I

Opposite page: Rangers on solitary patrol needed to pack all the supplies with them that would ensure survival and comfort for extended periods. Fred Tietge at headwaters of Muskeg River, 1948.

FRED TIETGE COLLECTION, COURTESY OF STAN TIETGE

brought my razor-sharp axe down on a block, it flew apart and I received a nasty cut above the ankle. I bled so much I was afraid that I had cut an artery. I pulled off my shoe and sock and grabbed the flour sack, heaping flour on the wound until the bleeding stopped. Then I bandaged it as best I could. Strangely enough I felt very little pain until I went to bed that night. I would drop off to sleep, and then it would grow numb and start to ache until I could find a comfortable position. This kept up for several nights. I limped slightly during the day and then it slowly healed, but the next winter it bothered me whenever my feet got cold.

Ted radioed headquarters and told them to either send a crew of white men or telephone Bob Holgate at the Muskeg District and have him dispatch the same Indians we had had on the Smoky fire. I preferred to have them, as they were now pretty well trained in firefighting. They would lose no time getting there and would bring their food along, such as it was. I was glad that's how it turned out, although a few extra came and I decided not to send them back. It was just as well I didn't, because this was a most stubborn fire and we needed all of them before it was finally stopped. I had asked Ted to fetch out extra supplies and he had brought along a Forestry horse, as his own horses were loaded.

As we had no chance to stop the fire from burning up the side of a mountain, we tried to prevent it coming down, as there was some vulnerable timber in the Sulphur River valley, besides being unsurpassed for scenery. With this in mind, we commenced making a fireguard as close to the fire and below it as we could. We had no thermometer, but I would hazard a guess that it was ninety degrees during the hottest part of the day from two to four p.m. The only way to buck the heat from the fire and the heavens was to eat our breakfast at sunrise and work in the cool of the morning and in the evening.

I staggered the shifts for the men who patrolled the fireguard, for if we neglected that chore, we could easily lose all our work in a few minutes. Several times a burning log would roll down the hill, jump the fireguard and start a new fire down below. Rolling rocks were the worst offenders as there was generally moss and pine needles sticking to them. I've seen some of them travel a quarter of a mile before stopping, leaving a trail of fire all the way. It was a great temptation for a man on patrol to pick out a shady spot and go to sleep. Ted and I had to be on the alert for this, as that might well have been the man's last sleep.

We managed to keep abreast of the fire and as we extended the fireguard, we backfired everything above it. Lack of sleep was getting everybody down. The saws, axes and mattocks were dull and it took double the effort to use them. Day after day the clouds got heavier and the temperature dropped a little, but it was still plenty warm.

It is a wonderful spectacle to watch a forest fire at night, especially on a mountainside. The flames ignite everything on the ground and creep up to a tree. It sizzles for a time until the pitch melts and catches fire. Then with a mighty roar the flames shoot up to the topmost branches and indulge in an orgy of destruction. This is the death knell and soon the tree falls with a mighty crash, sending a shower of sparks upwards, and the carnage goes on and on.

One of the Indians had brought his son along to do the cooking and keep the horses from straying. While cutting some meat, the lad's knife slipped and his hand was cut almost to the bone. His father saw it happen, but made no attempt to help the lad, who was forced to bandage it himself as best he could with a dirty rag. Ted came along shortly after and cleaned the wound and put a clean bandage on it, at the same time giving the father a great calling down, asking if he would want his son to get blood poisoning and die. Fortunately the lad had no more trouble with it. Those Indian children certainly learn their lessons in a hard school.

We had now reached a small swamp at the edge of a ravine and we were banking on checking the fire there. A small creek meandered through it, which would also help us, but should a spark start a major fire across the ravine our battle could well be lost. That evening we had a light shower, which raised our hopes, but it did not last long. Ted and I had reached the limit of our endurance by this time, so I put the crew on the night shift and told them that some of them could sleep, provided they didn't all sleep at the same time. They need not do any work, but were just to watch that the fire did not jump the ravine. We were so tired we did not feel like eating, so I just removed my boots and crawled into bed.

Around midnight I was awakened by a terrific crash of thunder, and thought if this turned out to be a dry thunder storm the whole country would be on fire. I did not have to worry long before the rain came – and how it rained! I jumped out of bed and got our firewood inside the tent, but I could not go back to sleep. A vivid flash of lightning would almost turn night into day, followed by peal after peal of thunder that sounded as though the heavens were being rent asunder.

I heard the Indians coming back to their teepees and wondered why there was a light burning in one of them every time I looked out. Towards morning, the thunder quit but the rain didn't, and I managed to sleep until an Indian came to our tent and asked to borrow some tea. I asked him if any of his friends were sick, as I had seen the light in the teepee all night. He laughed and said they were playing strip poker. It seemed they had started playing for money until such little wealth as they possessed was all in the pockets of a few. One young native had travelled from somewhere near Grande Prairie to Grande Cache and in a poker game he had lost his horse. He had been loaned a horse to ride to the fire and when he arrived he was quite respectably dressed. If the rain and the poker game had lasted any longer, he would have probably put on a Lady Godiva act minus the white horse. They had certainly taken him to the cleaners in that game, as all he still possessed was a pair of pants, an undershirt and a pair of worn-out moccasins. He didn't have a scrap of food, so Ted took pity on him and baked a couple of bannocks and gave him some tea.

As soon as it stopped raining, I headed for the Smoky River and took the Forestry horse along with me and my horses, as I wanted to pack a pump back

Looking down the Sulphur River from the end of Big Grave Flats, Athabasca Forest, c. 1916.

DOMINION FORESTRY BRANCH, ALBERTA FOREST HISTORY PHOTOGRAPHIC COLLECTION

to Grave Flats Cabin. Ted could cope with any little smokes that might show up, and I was anxious to see if the lightning had started anything around the Smoky. The following day I rowed across the river, crossed the Muddywater River on a couple of logs, and hiked a few miles down river. Trapper Faulk had a cabin not far from Stan Clark's boathouse. It was never locked, so I went in to see if he had a magazine I could borrow. I sat down and was leafing through one when I heard the sound of my horse bells across the river. Horses are very keen on salt and often came up to the cabin from the pasture. I always left salt out for just such occasions, so thought nothing more of it until I went to bed. Then I had a feeling something was wrong and could not sleep. Suddenly it came to me. Had those horses gone back to the pasture? I felt foolish asking such a silly question, as I had never had trouble with them there before, but I had to be convinced. I pulled on my pants and shoes and grabbed a flashlight and went out to the trail to look for tracks. Sure enough there they were headed back for Grave Flats.

I started out after them first thing in the morning. I had a big lunch along, and figured that surely the brutes would stop at Kvass Summit to graze. If I did not know how many crossings there were on that pesky creek before, I sure found out that day. There was no sign of the horses at the summit, so I ate my lunch there. I was certain they would stop at Grave Flats, but I would have to go to the fire camp and stay with Ted as I had no bedding at the Grave Flats Cabin and had taken all my food to the Smoky. This meant a twenty-five mile hike that day and I did a lot of searching before I found a log to cross the Sulphur River on. At that, I had to wade a narrow channel and crawl on my hands and knees over the log as the water was washing over and boiling underneath it. The bark was all stripped off and it was slippery as ice.

Fortunately, Ted was still at the fire camp. If he had left, it would have been the end of a not-too-perfect day for me. I was tired and thoroughly fed up with the Forest Service and ready to send in my resignation, but I thought I would sleep on it. As it transpired, it was quite a few years later before I finally resigned.

I found out later that there were only two places that the Forestry horse wanted to stay, at Rock Lake or Grave Flats. His former owner was an assistant ranger who committed suicide. Evidently he was not too ambitious and spent most of his time at Rock Lake or Grave Flats and his horses got so used to those places, they didn't want to stay anywhere else.

The Sheep Hunt

We had no more trouble with fires that year, so when the hunting season rolled around we decided to relax for a day or two and hunt the bighorn sheep. With that in mind we headed for the Summit Cabin, where the best hunting was. It's always wise to get out early in the morning when on a hunt, as you never know what difficulties may be encountered and how far you have to travel. It would be foolhardy to be on a mountain after dark.

Outfitter heading into the high country in the 1940s. The number of loaded pack horses suggests a long trip.

MARK TRUXLER COLLECTION, COURTESY OF JACKIE HANINGTON

We failed to locate any sheep until noon and then after we had searched a likely looking basin thoroughly with our binoculars, we spotted a small band lying down. Knowing they would likely be there for the rest of the day we planned our strategy. They were quite high, practically at the summit, so we would have to be careful or we would spook them and wind up with nothing. We had our choice of two routes to get above them. One looked easier, but we chose the other as the wind was in our favour. It took longer than we figured to get up on top. It is characteristic of that range to find an almost level plateau at the summit. We were pretty well winded on our climb, but decided to push on, thinking that the sheep might appear on top at any minute. As there was no cover we crawled stealthily on our hands and knees, peering cautiously ahead every once in so often so see if anything had showed up.

On one of these occasions I noticed a dark object which could very well be the horns of a ram, so I warned Ted and we lay flat on our stomachs deploring our luck. The object had not moved but I was positive it was not a rock, so we crawled some more. The wind was blowing quite briskly but was still in our favour. Taking another look as we drew closer, I recognized the object as a large golden eagle perched on the edge of the basin, gazing intently at the sheep. We crept up to within a few feet of him before he seemed to realize that he was in danger. He certainly was a surprised bird when he looked back at us. Then he blinked and hurriedly took off.

The sheep were now moving around, so we picked out two fair-sized rams and got both of them with our first shots. We bled them and removed their insides, then rolled them down the mountain with, of course, the hide left on. After an animal is dead, the flesh will not bruise. Needless to say, we had some delicious meals for a time.

Another Season on the Trail, 1934

That winter I completed my house at Vernon. I then rented it in the spring of 1934, and with my family motored back to Entrance. The mountains were calling me back. That was a lovely trip through the Okanogan, the State of Washington, up through the corner of Idaho, the Crowsnest Pass to Calgary and Edmonton, and home again. It was cherry blossom time and the orchards were a riot of colour. We would have liked to travel in more leisurely manner, but time was pressing. That season we rented the old Entrance railroad station, but it was not a bit comfortable, so we decided we needed to build a house of our own.

This summer also was hot and dry, and Ted Hammer and I kept our fingers crossed, dreading the thought of more fires. High water in the Sulphur River had washed out a section of the trail and we had not yet got around to fixing it. In the meantime, an Indian trapper had cut a narrow trail around it and at one spot two spruce trees were growing on a little knoll in a V-formation, with the trail passing between them.

About this time, our new supervisor, who was a quick-tempered sort of man, decided to make a tour of inspection with Ranger Harrison acting as guide. Most of the Indians have neat little packs for the simple reason, I suppose, that they don't have much to put in them, so this trapper had experienced no trouble passing between these two trees. The supervisor had his pack horse tied to his saddle horn and, suspecting no trouble, rode jauntily between them. As soon as the packhorse stepped over the knoll, the packs became lodged and he came to an abrupt stop. With the pull of the lead rope the saddle horse was snubbed suddenly, almost unseating the supervisor. During the time it took to extricate the horse – no easy task – the air was blue and Ernie said he expected a fire to start any minute.

The supervisor was under the impression that we still had a trail crew and why, he wanted to know, hadn't the trail been fixed. But, when Ernie explained that all maintenance had to be done by Ted and me in a district, which extended one hundred and fifty miles, he calmed down and the matter was never mentioned again.

Animal Stories

In most every district where hunters and sportsmen foregather, the talk generally centres on some animal that's in a class of its own and quite often grossly exaggerated. One case, for instance, was the grizzly I mentioned earlier with the enormous tracks that was later killed by Jack Brewster and Felix Plante.[42] Another case was an albino deer, which was in a bunch of four. I had been scanning the mountainside with my binoculars when I spotted him. He was a lovely animal, and probably his colour made him appear larger than he really was. As far as I know, no one else ever saw him and he was never taken, but I've heard guides mention the large tracks.

Then there was the Sulphur River Ram. He was seen and shot at often, but seemed to lead a charmed life. When Ted and I moved in to fix the trail, we camped close to a sulphur creek where we could heat water and have our baths. We had been told that there was a sheep lick close to there, so one evening we set out to find it. I had my camera along, just in case.

We could smell the lick while we were still some distance from it, so we sneaked up quietly. There were a number of ewes and rams there, but no spectacular heads. We watched them for awhile and then Ted nudged me. Some distance away, there stood the much-publicized Sulphur River Ram. He was a magnificent animal and we figured that he would measure at least nineteen or twenty inches at the base of the horns. The tips of the horns were badly frayed, but in spite of that he would still be a record head. It was easy to tell that he was suspicious as he kept glancing in every direction, and it took only the click of my camera shutter to stampede him as well as the others.

[42] Felix Fred Plante was born in Lac Ste. Anne in 1893. His family moved west in 1914 and he and his wife Caroline trapped in the Entrance, Big Berland and Obed areas. They had three sons and three daughters. They later moved to Jasper where he worked for 15 years as a guide for Fred Brewster. An F. Plante is listed as a forest ranger on the Athabasca Forest in 1925, likely Felix Plante. By 1927 Felix had saved enough to buy his own outfit that grew to 200 horses which he ran for 25 years. He was also noted for his hand-crafted pack saddles. Felix Creek was named after him. (Feddema-Jones 2007, Hart 1980, Karamitsanis 1991)

Jack Brewster and the Hawaiian Doctor

When hunting season opened that year, I happened to be at Eagles Nest Cabin when Jack Brewster arrived with a party of big game hunters. One of them was a doctor from the Hawaiian Islands who had never been near snow before and was looking forward to this experience. I told him that he might get tired of it before their trip was over. They were not too particular about filling their bag, but they did specially want moose, large heads if possible.

The day before, I had hiked from the Summit Cabin over to where the trail from Jasper Park entered the Forest Reserve and had put up a marker. Returning, I noticed three large moose on the trail and, as they appeared to want the right of way, I made a detour. It was close to the rutting season and they sometimes will attack a person at that time. I remember thinking that the largest one would make a fine trophy for some hunter. Brewster asked me if I had seen any moose lately so I told him about these three. He thanked me and rode on.

Three weeks later, I happened to be at Eagles Nest again when the hunting party returned. Brewster drew my attention to a large moose head on one of the packhorses. It had a spread of sixty inches, a record for that part of the country, and was one of the three I had told them about.

Hunters sought high ground to glass for game. Outfitter Mark Truxler in the mountains north of Rock Lake, 1940s.

MARK TRUXLER COLLECTION, COURTESY OF JACKIE HANINGTON

Fire in Rock Creek Valley

Nothing of any great importance happened in 1934 until one Sunday at the Summit Cabin. It was a sultry, hot day and the cabin was like a bake oven so we were leaning on the hitch rail outside gazing down Rock Creek Valley and wondering if the heat wave would end in a thunderstorm. We were not left long in doubt as a lightning flash accompanied by a loud peal of thunder broke the silence. A dry thunderstorm is always to be dreaded and this appeared to be just that. Suddenly we saw a bolt of lightning strike and in a few minutes a wisp of smoke rose lazily. The Summit Cabin was only about seven miles from the Jasper Park boundary, and we were not sure whether the smoke came from the park or our Forest Reserve, but as fire respects no boundaries we prepared to go to it.

It was late in the afternoon and I prepared some supper while Ted rounded up the horses, and we were soon on our way. It was dark when we arrived at our destination, so as a safety precaution we camped on the opposite side of Rock Creek to the fire. Nothing could be done at night, so we pitched our tent and went to bed. A slight breeze sprang up during the night, but there was no hint of rain and early next morning when we got to the fire it had already burned about three-quarters of an acre of pine. It was easy to tell where it had started as the lightning had driven a sliver from a dead tree and it was stuck in the ground about twenty feet away. The tinder-dry moss had ignited and a forest fire was in the making.

We worked like Trojans to get a fireguard around it and at times it looked like a losing battle. A spark would jump the guard and if not extinguished would spread amazingly fast. At last we had it surrounded and could breathe a little easier, but were both tired and hungry as we had not dared to stop for lunch and it was now well into the afternoon. We took turns going down to our camp to eat.

Earlier in the day, a plane droned lazily overhead but we paid little attention to it as the government plane had been engaged in aerial photography. I still thought it funny that they would be up with so much smoke in the air.

The next day we had it much easier as it was just a case of watching that nothing hot got over the guard, while the fire gradually burned itself out. While one of us watched, the other fished in Rock Creek. That night there was every indication of rain and I cut a good supply of dry wood. It was just as well that I did, for towards morning the rain amounted to almost a cloudburst and I had to have Ted's assistance to kindle a fire. He held a canvas over me and I was able to keep it going.

The time dragged horribly as we did not feel like going anywhere and getting our clothes wet, as it would be impossible to dry them. We had no stove in the tent and all our cooking had to be done on an open fire. I had brought along a couple of magazines for just this sort of eventuality and had read them through and through, including the advertisements, when the rain finally abated.

A Service to Jasper National Park

When the rain stopped the first thing we had to do was to ascertain just where the boundary was between the Forest Reserve and Jasper Park. There was little or no line cut through the timber in the mountainous part, so we climbed a mountain that had a stone cairn as a marker on its highest peak and, by sighting across to a neighbouring peak, we soon found that the recent fire was in the Park.

Resting while on our climb up the mountain, we were treated to a magnificent sight. A large buck deer stepped out on a ledge and gazed across the valley. Monarch of all he surveyed, so to speak. It looked almost as though he was posing for our benefit for he knew perfectly well of our presence. He wore the loveliest pair of antlers that it has ever been my good fortune to see. With all his masculinity, he also had a dainty manner. When a bothersome insect annoyed him he would stamp one foot and nip the offender. His ears were constantly twitching and when he decided to leave that spot, a few graceful jumps took him out of our sight.

As I neared Rock Lake, I noticed the smell of burnt wood and vegetation usually associated with forest fires, and a short distance from there I rode into the greatest scene of devastation I have ever witnessed. The trail had been made passable by firefighters and as I rode along, I wondered if our cabin was still standing and what damage had been done. All this was totally unexpected, as we had been unable to see any smoke from where we fought the fire in the Park. The high mountain range that blocked our view could only be traversed by way of the Mountain Trail through Eagles Nest Pass. I wondered if my radio was destroyed, too, as I had left it at Rock Lake. I had not been packing my radio lately as there was no use both of us doing so as long as we were together. Ted had not bothered to pack his to the fire as we only had a small tent, and if anything serious had happened one of us could have ridden back to the Summit Cabin. I was happy to find that although all the timber was burned around it, the cabin still stood. Better still, no damage had been done across the lake where the scenery was beautiful.

I was just unpacking when a park warden rode up and gave me the details. This fire had been started by lightning, and our lookout man had spotted the smoke, but couldn't be sure whether it was in the Forest Reserve or in Jasper Park. The Park people were notified anyway, and the superintendent immediately chartered a plane and flew over the area. This was the plane that I heard overhead at the Park fire. Once he had assured himself that the fire was in the Forest Reserve, he lost no time in getting back to Jasper and sending a report to our supervisor. In it he stated that he had looked in vain for a sign of forest rangers or firefighters in the vicinity of the fire, but had dispatched some of his wardens and a crew of firefighters with orders to stand by in case the fire got into the park.

Naturally, our supervisor called for an investigation and I found a note in the cabin instructing me to report to Headquarters without delay. A crew from

Eagles Nest Pass between Rock Creek and Wildhay River, around 1920. The eagles nested in a shallow cave on the face of the low cliff above the pass.

DOMINION FORESTRY BRANCH, ALBERTA FOREST HISTORY PHOTOGRAPHIC COLLECTION

Entrance had been dispatched as soon as the fire was reported and had arrived at Rock Lake the same day as the superintendent of Jasper Park had been joyriding in the plane. One of the Forestry crew rode down to the cabin that evening, as they were worried about our safety. Their camp was several miles away on the Mountain Trail, and I would have seen it if I hadn't taken the other trail. He said there was little danger of the fire spreading as they were watching it closely and that any hot spots remaining were well inside.

I told him that they might as well stay for a few days more and that I would be back from Headquarters just as soon as possible. When I arrived in Entrance, the supervisor was already there. When I told him what had happened, he praised us for our prompt action and said that he was looking forward to meeting with the Park superintendent. I can well imagine what transpired there.

As soon as I got back to the fire camp, I sent the crew home as the fire was practically out and the weather looked unsettled. I spent a few days estimating the damage and mapping the burned area before proceeding upcountry. I would be finishing the season alone as it was getting late and there was nobody available to take Ted's place.

Hunting season had rolled around again and I had gone out after goat on the Labour Day holiday in the Eagles Nest area. I was not too enthusiastic, as sheep hunting was more in my line, but I had been lonesome since Ted left. I spotted a nice looking billy and, although it was a long shot, I thought I might as well try my luck. Taking careful aim I pulled the trigger and the only response was a click. I was using a lever action .30 calibre U.S. long rifle, so I pulled back the hammer and tried again with the same result. By this time the goat had disappeared and I left in disgust, also thanking my lucky stars that I was shooting at a goat instead of a grizzly bear. I ejected the dud shell and put it in my pocket.

Fred Wharton's Story

Sometime later I was at Rock Lake Cabin. It was evening and I was lying in my bunk reading a magazine. I had baked bread that day and as the cabin was hot, I had the door open. I heard a sound outside, but paid no attention as I figured it was likely a horse or a deer at the salt block. There had been sharp frosts at night, so I kept the slop pail inside by the door to prevent it from freezing.

A sound by the door made me take notice, and what I saw made my hair stand on end. The slop pail was in the shadow cast by the door and, although indistinct, I could discern the shape of a head and the shoulders of a dark-coloured animal busily engaged in rummaging through the contents of the pail. All types of animals flashed through my mind, but none seemed to fit. Wolf, bear, wolverine, dog or coyote were all promptly ruled out, and I was reaching for my rifle when a voice said, "Get your head out of there, Blackie and come and get unpacked."

Now the mystery was really deepening, so I lost no time in getting outside, and I beheld a strange sight. In the light cast from the cabin window I saw a tall man unpacking the smallest Shetland pony I had ever seen. Nearby were three more pygmy-sized ponies.

I introduced myself, and the gentleman in return told me that his name was Francis Wharton. I told him to finish his unpacking and come in and have supper.

Wharton was indeed a most interesting person. He could be termed a gun crank, but he did not confine his skill to firearms. Repairs for watches, cameras and many other delicate instruments were no problem to this ingenious man and he made his repairs for a ridiculously small fee.

After supper I told him to fetch in his bedroll and sleep in the other bunk. He thanked me and did so, and when he had prepared his bed he commenced to work on a coyote hide that he told me he had shot just about an hour before. I thought this must be an exaggeration, as it had been dark for about three hours and a coyote presents a difficult target in bright sunshine. He must have sensed that I doubted his word so he went on to explain that when he heard the coyote howl, he took his rifle and tried to locate it through his telescope sight. A full moon and a light powdering of snow helped him in his search and finally he was able to centre the crosshairs of the scope sight on the coyote, which was standing on a knoll howling at the moon. I was now fully convinced, after I felt the hide and noticed a slight trace of heat still in it.

We talked well into the small hours of the morning and I'm ashamed to say that I fell asleep. I told him about the goat hunt and the dud cartridge and he became interested right away and asked if he might have it, as he wanted to find out why it had not functioned. I asked him how he could find this out, and he explained that he would split it down the centre and examine the percussion cap first, and so on. During my service in the army I had become fully aware of the terrific power of fulminate of mercury enclosed in the caps, so I remarked that would be a rather dangerous practice. He grinned and said yes that was so if one didn't know just how to go about it.

Having recently bought a new riding saddle and not wishing to use neatsfoot oil or a grease that would darken the leather, I asked him if I could have the fat from the carcass of the coyote as I had been told that when rendered out it was perfectly transparent and was excellent when used on leather. He said that he would be passing the carcass the next day and would keep that in mind.

I got to know Wharton very well as he camped near us at Entrance for some time. He owned a heavy old Cadillac car with which he pulled a trailer that housed his four little Shetlands, which he claimed were the smallest in the world. Shortly after his visit to Rock Lake, I met a halfbreed on the trail who told me, "I see funny 'ting my fren'. I see one heap big man track and maybe three-four little colt track – me no savvy dat 'ting!" When I explained it to him, he laughed heartily and shook his head. "White man crazy for sure," he was probably thinking.

I lost track of Wharton for several years until I happened to run across an article in a newspaper telling of a Mr. Francis Wharton, a resident of Little Fort, B.C., who had been having trouble with his teeth and visited a dentist. After the examination, the dentist told him it would be advisable to have his teeth extracted and replaced with false teeth. Upon being quoted the price, Wharton figured it was beyond his modest means, but told the dentist to extract the teeth since that was something he couldn't very well do himself. After his gums hardened, this resourceful man figured out a way to install his own plates. He hunted and killed a young buck deer and carefully removed its teeth. Then he made an impression of his gums in plastic wood and painstakingly inserted the deer's teeth. With this accomplished, he could eat his venison in comfort. He did not live long enough to enjoy his new dentures as it was announced over the radio that he had died soon after.

At the end of 1941, the Headquarters ranger L.L. Waikle, resigned, and since I was next in seniority I took over. It was wonderful to be home most of the time, though I missed the free-roving life that I had grown accustomed to over the years.

On Patrol from Entrance, Jack Glen looking forward to another tour of duty, c. 1940s.

JACK GLEN PHOTOS, ALBERTA FOREST HISTORY PHOTOGRAPHIC COLLECTION

CHAPTER NINE

LAST TRAILS
1935–1945

Family Life in Entrance

In the fall of 1936 I decided to build my own house, as I could not afford to pay rent, especially now that I was laid off during the winter. Providing that I did not use green timber, the Forestry people allowed me all the dead timber I required, both for building and for firewood, and also the use of a team and sleigh. This was an ideal arrangement as much of the timber, although dead on the stump and sometimes blown down, was surprisingly sound and dry. If you build with green timber, the walls remain damp and settle considerably in the drying out process. Sometimes during cold weather the logs will check and crack like a pistol shot.

My wife peeled off the bark while I hewed the logs on the inside and raised them. I was fortunate enough to get lumber quite cheaply at the abandoned mining town of Pocahontas. The shingles or shakes I made myself. After the landscaping was finished, we were quite proud of our new home, and years later we hated to leave it. The dangerous part was the location as far as our son was concerned, as we had built on the banks of the swift-flowing Athabasca River. However he heeded our warnings and played with the neighbour's children clear of the river.

Harold Lake and Fred Hendrickson came back to Entrance and I was instrumental in getting them the traplines that they wanted, Fred in the Berland River District and Harold at Sheep Creek.

Jack Glen and his wife built this home for their family at Entrance in 1936.

JACK GLEN PHOTOS, ALBERTA FOREST HISTORY PHOTOGRAPHIC COLLECTION

National Forestry Program, 1938

In the fall and early winter of 1938 I had a number of National Forestry Program boys under my care. I taught them woodcraft, firefighting, care of animals and in fact anything that would be beneficial to them in the line of forestry work. They also built eight miles of new telephone line from Winter Creek to the Wildhay River Cabin. We carried on this work until it was too cold to live in tents and then they were disbanded.

Wildhay River forestry cabin in the 1950s.
COURTESY OF JACKIE HANINGTON

The War Years

The following year (1939) war was declared and we were sworn to secrecy in the use of radios. One day on the trail I met Stan Clark with a wealthy Swiss baron out for an extended hunt. As usual, very few hunting parties ever bothered packing radios and when Stan asked me how everything was going in the outside world, I told him that we were at war with Germany. When the baron heard this, he was quite upset and said to Stan they must turn back as he wished to cable orders to his brokers in England. Otherwise, he said, he stood a good chance of being ruined.

On this point I saw a chance of repaying Stan for his many kindnesses to me and also let the hunter get on with his hunt if I could radio his message for him. It would mean a financial loss for Stan if the party had to turn back. When I proposed this to them, all I asked was that no military secrets be included in the cablegram. The baron (whose name I have forgotten) was quite relieved at this, and promised that nothing detrimental to the Allied Forces would be entered in the communication. It would be in code difficult to decipher without a key. I had met them not far from Grave Flats Cabin so I turned back with them, and they made camp there and waited until time for my evening call-in schedule. That

National Forestry Program

"Terk" Bailey (l) and Stan Hughes, National Forestry Program Field Office, Jasper National Park, 1939. Bailey later became Deputy Minister of Lands and Forests in Ontario. Hughes served with the RCAF in WW II, joined the AFS 1949 and retired as Director of Forest Protection in 1975.

STAN HUGHES PHOTOS, ALBERTA FOREST HISTORY PHOTOGRAPHIC COLLECTION

The onset of the Great Depression hit Alberta early and hard. Programs to assist unemployed men began as early as 1930. Unemployment Relief Camps (URCs) were set up by the Province, running until 1937. The Dominion government started its own camps in 1934, growing to 120 across Canada. The camps were evidently of a make-work nature, with harsh rules which led to ahigh turnover and great discontent that culminated in riots in 1935. The Dominion program ended soon after. Alberta's Youth Forestry Training Program (YFTP) supplanted the URCs in 1937, receiving 40 percent funding from the federal government. The program consisted of camp-based training in basic forestry and work on projects such as trail, road and bridge construction. In 1938 the camp-based program continued, but another approach involved assigning selected men to Forests to work with rangers as assistants. In 1938, 35 men worked out of the Highwood River Camp in the Bow River Forest, and 25 YFTP trainee assistants were sent out – twelve to the Crowsnest Forest, four to the Clearwater and nine to the Brazeau-Athabasca. In 1939 the fully Dominion-funded National Forestry Program (NFP) replaced the YFTP, using a similar combination of project-based camps for basic forestry training as well as trainee assignments to rangers as assistants. The Brazeau-Athabasca Forest was assigned one 10-man unit to work in a camp and an additional ten "trainee assistant rangers." By the end of 1940, the program ended as the war effort increased. Many of these trainees who went to war returned to continue careers in forestry upon their return after 1945.

message was the most meaningless jumble of words that I had ever seen, and to make matters worse, reception was poor. After I had it read back to me several times until it was correct, I was all of a lather. First I radioed it to the Headquarters ranger, who in turn phoned it to the station agent at Entrance. He wired it to a brokerage firm in Toronto who cabled London. Reception was much improved the following morning and the answer came through to the effect that everything was OK and instruction had been carried out. I felt like a boy scout who had done his good deed.

Ted Hammer Leaves the Athabasca Forest, 1940

Ted Hammer had applied for a timber inspector's job,[43] so this would be our last season together. As usual, we headed for Rock Lake where we would have to wait for the grass at the higher altitudes to make horse feed.

[43] Ted Hammer was promoted to timber inspector and moved to Fort McMurray in fall 1940.

The ice in the lake was breaking up and, as I was fish hungry, I suggested to Ted that we fix up Stan Clark's boat and go fishing between the ice floes. Stan had hauled out an old boat during the winter, and we certainly had a lot of enjoyment out of it. We kept it in good repair to show our appreciation. Ted told me I ought to have my head examined as any self-respecting fish would be hugging the stove on a chilly day such as that was. However, I pointed out that the kingfishers and ospreys had more ambition than he had and seemed to be doing all right, so he reluctantly gave in and we shoved off, with him rowing.

We had just got out to deep water when I had the strike. My reel started to scream and the inevitable happened and I could not do a thing about it.The line snapped, and all Ted had to say was, "I wonder when the Loch Ness Monster moved into Rock Lake." We changed places, and soon Ted had a fighting fool hooked. He played him until he thought he had him licked, but the fish was just playing possum. As soon as Ted got him close to the boat, the fish began fighting again. These Lakers have a nasty habit of standing on their tails, turning a somersault and spitting out the hook. I was just about as excited as Ted was with that fish and had allowed the boat to drift quite close to the shore. I suddenly got an idea. "Hold everything," I said. "I'll beach this landing craft," and I picked out the sandiest part of the beach I could find and drove the boat onto it. Ted made a flying leap and landed his fish clear of the water without any more trouble. We guessed its weight as about eight pounds.

Ted went back to the Summit Cabin to pick up his belongings as his application for a timber inspector's position had been accepted. I headed for Rock Lake over a trail which I had never before travelled. In the years that followed, I missed Ted, as nobody could wish for a better trail partner. We had been through thick and thin together with never a discordant word to mar our friendship.

Archie Clark, the game guardian, was an ardent fisherman too, and one time we three met at Rock Lake. Archie had been fishing and had quite a nice lot of fish, of which he was naturally quite proud. Ted and I decided to have some fun

Jack Glen and Ted Hammer at Rock Lake Cabin, c. 1939.

JACK GLEN PHOTOS, ALBERTA FOREST HISTORY PHOTOGRAPHIC COLLECTION

with him, so said we expected something better as those fish of his were the size we used for bait. "All right, you smart alecs," he said, "I'll show you some real fish one of these days." He was as good as his word. Ted had been to Entrance and on the way back, met Archie on the trail. "When you get back, take a look on the top rail of the corral and you'll see something worthwhile," Archie told him. Ted was puzzled but when he got to the cabin, lo and behold, there was a huge fish head on the corral rail, smelling to high heaven. He left it there for me to see also. I've seen many large salmon taken from B.C. rivers, but none of them equalled Archie's lake trout as far as the head was concerned, anyway. We were all ready to accuse Archie of buying the fish from some fishmonger, but we never had the chance as he was transferred shortly after that.

The Hockey Player

Later that spring I had a new partner, Irvine Frew. He had been a professional hockey player with the Montreal Maroons and was a keen horseman and ardent fisherman.[44] As he didn't report to Headquarters until late spring, we missed the sport in Rock Lake that Ted and I had enjoyed so much. This was when the ice in the lake broke up and the large lake trout emerged from the depths. I managed to hook one or two whenever I felt lonesome or had a yearning for fish.

One day I rowed the length of the lake and trolled without success. I was reeling in my line when I noticed a fair-sized trout tagging along behind my spoon and lure. Someone had told me that the tip of a squirrel's tail tied on a hook made an excellent lure, but as I did not want to kill a squirrel just for that, and as mountain gophers were plentiful and also a nuisance, I killed one of them. These gophers used to burrow under the cabins and sometimes made the night hideous with their quarrels, love-making or what have you. On washdays I used to empty the soapy water down one of what seemed to be a maze of tunnels. Presently one or more of them emerged very much bedraggled and seemingly very perturbed, but they never took the hint and moved out.

To return to my fishing: the fish tagged along about six feet behind the gopher tip lure, but made no attempt to take it and appeared merely curious. As I was now in shallow water and amongst weeds I had to reel in to avoid getting my tackle fouled. That fish got quite close to the boat before he noticed me and then beat a hasty retreat.

Irvine Frew and I were on our way to attend Ranger School, which was at Jasper that year, and we had reached Rock Lake. As usual after supper our thoughts turned to fishing. It was a lovely evening and the lake was as smooth as glass. The fish were rising well and nothing marred the peace of the evening except a loon uttering his ghoulish call. We could faintly hear the tinkling of our horse bells in the direction of Wildhay River, always a welcome sound. I rowed the boat that evening while Irvine fished. We soon had all we wanted, and Irvine suggested that I quit rowing while he played the last one who appeared to be quite a fighter.

[44] Irvine Bell (Irv) Frew had a 37-year career as a forest ranger with the Alberta Forest Service from 1934 to 1971,mainly based on the Bow River Forest. He moved to the Athabasca Forest in 1939 and transferred back to the Bow River late season 1940. His hockey career spanned the period 1926/27 (Calgary Tigers, PrHL) to 1940/41 (St. Louis Flyers, AHA). During this time he played three seasons in the NHL: Montreal Maroons 33/34, St. Louis Flyers 34/35, and Montreal Canadians 35/36, earning a total of seven points with two goals and three assists. Sources: HockeyDB.com. The Internet Hockey Database; Employment records, Government of Alberta.

As the boat stopped drifting, the fish tired somewhat, but not enough to stop him slowly towing the boat. We let him have his way for a time and then gaffed him and slowly headed for the shore. But our sport was not over yet. Suddenly, Irvine drew my attention to a school of Rocky Mountain whitefish in fairly shallow water of about seven or eight feet. He said that these could be caught but it called for a different technique.

He went back to the cabin and rummaged through his fishing kit until he found several small fly hooks. For bait we used the remains of a can of corn that we had for supper. We rowed around until he found another school and Irvine baited the hook generously. They were quite hungry or else the corn appealed to them for they jostled each other out of the way to get at it. The mouth of a whitefish is tender, so Irvine took the line off his rod and lowered the bait over the side of the boat by hand. In the first attempt the bait was knocked off the hook but the second try proved fruitful. The fish darted hither and yon and Irvine let him have his head until he tired before gently pulling him into the boat.

Next morning we left Rock Lake at daybreak for Entrance. It was considered a two-day trip, but doing it in one day gave us an extra day at home. This was the beginning of the Ranger Schools, which have since become a yearly event, and it was held at Jasper with the park wardens attending. It was a most enjoyable affair and by swapping one's experiences with others, a vast amount of valuable information was gained.

Entrance Ranger station (l), stables and corral on flats at Entrance, Ted Hammer's cabin above them on the hill. The Canadian Northern rail line is seen between the buildings and the Athabasca River. Across the river, the Grand Trunk Pacific rail line is clearly visible. The log cabin by the Athabasca River is probably Blackie Langelle's, c. 1920s.

JACK GLEN PHOTOS, ALBERTA FOREST HISTORY PHOTOGRAPHIC COLLECTION

Many new methods of firefighting were discussed and a number of different fire pumps had now appeared on the market – some good others not so good. Woodsmen were also becoming fire conscious, and were doing their best to educate the campers and hunters. Although it was a far cry from the present highly organized method of firefighting, it was a step in the right direction and the improvement has been quite noticeable as time has gone by. I attended schools in Edson, Edmonton and Calgary before resigning from the Forest Service.

Serious Infection on Patrol

Irvine Frew did not finish the 1940 season in my district as he was transferred to southern Alberta and while I was alone I had a freak accident that could have been serious.

The horses sometimes wandered a long way up the Wildhay River, partly to get to their favourite gravel bar where they stood and whisked flies from one another, and also in search of the succulent goose grass that thrives in shady, damp spots along the creeks and rivers. On this particular occasion, they had gone further than usual when I caught up to them. Not wishing to get my feet wet I had stayed on one side of the river all the way, although I was tempted several times to wade across when the walking looked better on the other side.

I rode bareback to the cabin, packed up and headed for Eagles Nest. Just before dropping off to sleep I felt something crawling on my leg so I got out of bed, lit a candle and investigated. Four or five wood ticks were crawling about aimlessly while one was in business. He had evidently tried two other spots as was evident by purple blotches about the size of a twenty-five cent piece.

I yanked it off, caught the others, put them in a paper bag and cremated them in the stove. I found no more in my clothes so I gave the matter no further thought. It was easy to figure how they got there. The moose harbour them and if a horse lies down where moose have been they generally pick up a few. My saddle horse must have passed them on to me. I have seen moose so badly infested with ticks in the early spring that it would be a deed of mercy to put them out of their misery, which I have often done.

About six weeks later my leg started to itch and to pain horribly on the spot where I had been bitten. I was camped between Big Grave Flats and Grande Cache, about eighty-five miles from home. This is a mighty long way especially when it is torture to either walk or ride. I notified the Headquarters ranger by radio and asked him to stand by as I might have to have a plane land on Rock Lake and pick me up if my condition got worse. Harry Hammer,[45] Ted's brother, was the closest ranger to me, and he begged me to stay where I was until he could get there and help me. I thanked him but figured I had better not delay as I had visions of having my leg amputated.

[45] Harry Hammer joined the Dominion Forestry Branch in 1926 and worked as a forest ranger on the Athabasca Forest to 1946, later serving seasonally as lookoutman.

I hobbled out and caught my horses and reached Grave Flats Cabin on the first stage of my journey. I went to bed that night but had little comfort, as I had nightmares and was probably delirious part of the time. I tried to think of something that might ease the pain and decided on a fat bacon poultice, which helped enough to allow me to sleep for probably two hours. In the morning when I took the poultice off, it was covered with pus and something that resembled black ink.

I decided to double up on my travel even if it was agony and bypass the Summit Cabin. The cabin was about a mile from the Mountain Trail and when we came opposite it, the horses could not understand why I was going on, as they had been in the habit of turning in there no matter what direction we came from. Nearing Eagles Nest Cabin I was pleased to see that a hunting party was camped not far away, and as soon as I unpacked I limped over there. It was one of Stan Clark's parties and my neighbour Cliff Faulk was the cook. He cooked a hearty meal for me and then dressed my leg. It was torture to touch it so I objected when he suggested putting a boiling hot bottleneck over it. That evening Mark Truxler came over to the cabin and offered to go all the way to Entrance with me if I thought it necessary. He was another of my neighbours, but I knew his services as a guide for the party would be missed if he went with me. I told him if he could come over the next morning, dress my leg and come as far as Rock Lake, I would decide then whether to fetch a plane or continue with the horses. This was only a twelve-mile journey and not too strenuous. My leg pain was less, so I decided to go on from Rock Lake alone. The wound was on the inside of my leg so I could not ride naturally, but had to have it hooked over the cantle of the saddle. It cramped often, that meant a stretch of walking, or rather limping. My greatest fear was of falling asleep and tumbling out of the saddle, but at long last I reached Entrance.

My wife didn't drive and Jim was still too young, so we started for Jasper by train the following day and made it all right. Kindly old Dr. O'Hagan took one look at my leg and ordered me into the hospital right away. He shook his head and

Mark Truxler (l) and hunter Paul Schmidt with moose kill, 1940s. Mark was an enthusiastic guide, outdoorsman, and the father of Jackie Hanington who contributed many of his photos to this book.

MARK TRUXLER COLLECTION, COURTESY OF JACKIE HANINGTON

Mark Truxler

Mark Truxler was born in British Columbia 1905. He went to Banff in 1923 and Jasper in 1927 to work as a guide. In 1930 he married Agnes Harragin who, with her sister Mona, were the two first women to be licensed guides in Jasper. The Truxlers moved to Entrance in 1936 where Mark continued to work as a guide while they raised their family. He worked at Miette Hot Springs from 1950 to 1955, then became supervisor of the East Park Gate until he retired in 1970. Mark and Agnes were avid historians and collectors of local history artefacts and memoirs. They had two children—Vernon became a timber cruiser for North Western Pulp & Power in 1955, and Jacqueline married Bill Hanington, an Alberta forest ranger at Muskeg and Cabin Creek before joining NWP&P. Jackie contributed several of Mark's pictures which are used in this book.

(Feddema-Jones 2007, Hart 1980.)

told me that I had just got there in time, for had gangrene set in I would probably have lost my leg.

The Jasper hospital was just like a home away from home. With the care I received from Dr. O'Hagan and the staff I was able to go home in three weeks time. I could possibly have saved myself all this fuss and pain had I applied something very hot to the body of the wood tick instead of just knocking him off and leaving the barb embedded in the flesh. Heat would have caused it to withdraw the barb. From the hospital bed I had an uninterrupted view of the mountains in all their grandeur and I wondered if in my weakened condition I would still be able to hunt the elusive bighorn. I soon regained my strength when I got back on the trail, and did succeed in bagging one.

Dr. Thomas O'Hagan

Dr. Thomas O'Hagan set up practice in Jasper in 1924 and served the community for over thirty years. He founded the Seton Hospital in Jasper. In the early years, Hinton did not have its own hospital and patients had to travel to Jasper or Edson by train. If no train was scheduled, this trip was often made by railway speeder.

Source: Hazel Hart, *History of Hinton*, Friesen, Edmonton, 1980. p. 334.

The First School at Entrance

Education for the children of our little community was a problem over the years. We had hired a high school graduate to tutor Jim, but soon realized that this method was not going to prove satisfactory. Marie was a wonderful girl and outside of classroom sessions she and Jim were the best of pals. He would study all right for a short time and then he kicked over the traces, so we decided something had to be done. We could have the services of a provincial teacher, providing that we had nine pupils and a place to teach them.

The cabin that Ted Hammer and his family had occupied was the only available place for a school, so when we approached the Forestry people and they kindly let us have the use of it, providing that we did not ask them for the cost of any alterations. This presented no real problem as Fred Hendrickson volunteered to help me with the work. We cut practically all the logs out of one side of the building and put in windows to provide sufficient light. The mining town of Brûlé was now abandoned and just another ghost town, so we obtained the desks from that school.

Entrance headquarters, 1940s. Ted Hammer's house, converted to a schoolhouse, is on the hill above.

ERNEST HARRISON COLLECTION, COURTESY OF LUCY MCRAE

Entrance School Children early 1940s, l-r Alan Moberly, Jim Sibleyk, Mabel Moberly, Jim Glen, Vern Truxler, Eileen Chapman, Jackie Truxler. Jackie relates that they were enjoying some watermelon, a rare and unusual treat for those times.

MARK TRUXLER COLLECTION,
COURTESY OF JACKIE HANINGTON

This arrangement worked fine until new families started to move into Entrance and the surrounding district. Some of them had children and the school problem again became acute, so we looked for another solution.

We were fortunate in having a parcel of land in the new community of Entrance deeded to us free, and all the lumber we required donated by the mining company at Brûlé. We had to salvage it ourselves, of course. Outside of a carpenter's wage, we were little out of pocket. A teacherage was also built and we were again in business.

Fire at the Drinnan Mine

For a change of work once in a while we had the usual hunter's fire to put out. There was also a fire that started at a small coal mine at Drinnan, not far from the site of the present large pulp mill at Hinton. This fire could have been serious as it was close to the mine buildings. Ranger Gordon Watt[46] was the only available man and unless the two of us could not handle a fire, the supervisor was loath to hire a crew when the fire season was officially over.

This evidently was a wet mine as they had a pump working continuously and the water was allowed to run down a slight incline where it seemed to seep away in the rocks and gravel. The mine supplied the north country or a great part of it with coal. The coal was double sacked, hauled to Edmonton, and then flown to its destination.

At the time of the fire, only two miners were working, one on the day shift and the other on the night. Our water supply was solved by diverting the water from the mine into a large tank and pumping it from there onto the fire. All we could do was keep the fire from spreading as we didn't have time to dig any fireguard. This meant we had to keep circling the fire constantly to keep the margin damp. We ate in shifts also as it would no doubt have spread if left unattended. We generally started pumping about eight o'clock in the morning and quit at six in the evening, but watched it until bedtime in case the wind sprang up. After two days the area in the centre had pretty well burned out. A good soaking rain finally put it out.

Patrols and Bear Encounters, 1941

The following spring (1941) I had a new partner, Clarence Earl.[47] He was a jolly fellow and very fond of fishing. Our two favourite spots were Rock Lake and the little lake close to the Summit Cabin. One Sunday afternoon, Clarence and I were fishing in the little lake almost at timberline when we noticed movement in the buck brush. Pretty soon a silver tip three-year-old grizzly bear came into view. A very old bear, or a three-year-old bear, are in my estimation the most dangerous. This particular bear was hunting gophers rather than digging them. He would make a mighty bound, and once in awhile we could hear the death squeak of a gopher.

[46] Andrew Gordon Watt worked as a forest ranger on the Athabasca Forest from 1942 to 1952, as recorded in government reports. Hart (1980) states that he started in 1940 and became chief ranger in 1943 — he may have been a fire ranger on wages from 1940. He bought the Entrance Store from Stan Clark in 1951, which became legendary for its old-time general store atmosphere and hand-operated gas pumps. (Feddema-Jones 2007, Hart 1980)

[47] Clarence Ethbert Earl was a forest ranger in Alberta from 1926 to 1944 and was on the Athabasca Forest at least in 1941.

This was the time we should have made our exit, but we were so interested in Junior's antics that we stayed. Pretty soon he strolled over to the lake for a drink. Then he seemed to pick up our tracks and followed them towards us. "Isn't it time we were getting out of here?" Clarence asked. "No," I replied, "It's foolish to run from a bear if you have no tree to climb, which we don't have."

"Well, what are we going to do, just stand here until he makes up his mind which one of us to start on first?" he asked. "We must try to scare him," I said and I pulled off my old red mackinaw coat and swung it around my head. Clarence started cursing in a rich Irish brogue, which I'm sure must have caused some of his ancestors to turn over in their graves, but Junior still kept coming. He reared up on his hind legs, took a good look at us as much as to say, "This ought to be easy picking for me," and came on again. By this time things had reached a serious stage, so as a last resort I started moving towards him (not fast, I'll admit), and when I was close enough I threw the old mackinaw at him and drew my hunting knife. That did it. He turned and fled. We made a solemn vow then that we would never go fishing without a rifle in the future.

That season was cold and wet so we were not plagued with fires. While Clarence patrolled the Sheep Creek country, I set up a snug little camp at Little Grave Flats with the intention of cutting the brush, which was encroaching on the trail. I wore chaps when riding, but some other people were not so fortunate, and the result was that they got drenched. However, I only got in one day's brush cutting when I was taken sick. My appetite left me, which was something unusual, and I couldn't sleep. The only things that appealed to me were canned tomatoes, toast, tea and coffee. When my limited supply of canned tomatoes came to an end, I really felt miserable. The bread played out next and as I didn't have ambition enough to bake more, I made hot cakes. I always carried a small first aid kit, but it consisted mostly of bandages and salves. However, there was a small bottle of castor oil that I had not noticed before. Much as I detest the vile stuff, I drained it to the last dregs. Next morning I felt like a new man and my appetite had improved, but it was time to head back to Entrance.

I tried to keep a pretty strict schedule on the day I was due home, as my wife and Jim were always out to meet me. As soon as I hoisted Jim into the saddle he would ask, "Did you catch any fish, Daddy?" and I could generally assure him that I had. There was a nice stretch of water on Solomon Creek near Major Fred Brewster's roundup cabin where I used to stay if it was not occupied. It was always good for half a dozen or so brook trout.

Late that fall I was headed for Grave Flats Cabin. It was my last trip of the season and I had come from the Walton Creek Trail and a chilly wind was blowing, so I dismounted and led my saddle horse to warm up a bit. The trail here followed the contour of a mountain above the Sulphur River and was just wide enough to accommodate a packhorse. I was ravenously hungry and was trying to decide what I should cook for supper when my saddle horse stopped and snorted.

Jack Glen and son Jim at his Entrance home, in front of the log garage and woodshed he built, late 1930s.

JACK GLEN PHOTOS, ALBERTA FOREST HISTORY PHOTOGRAPHIC COLLECTION

Opposite page: Map 6. Trails and Cabins in the Athabasca Forest, 1945.

When Jack Glen built the trail (1924) from the Wildhay River to Donald Flats, he named it the Donald Flats Trail; when he built the trail down the Berland River (1925) to the 15th baseline he named it the Berland River Trail. On the 1945 AFS map of the Athabasca Forest, the names are reversed. Neither trail remains intact today. The 1945 map also refers to the stretch of trail from the Smoky River to Sheep Creek as the "Dry Canyon" trail. Currently, it is again called the "Sheep Creek Trail" as originally named by Jack Glen when he built it in 1927.

FOOTHILLS RESEARCH INSTITUTE HISTORY SERIES

Lo and behold, I was face to face with a grizzly. We saw each other at the same time, and he was just as much surprised as I was. Letting out a startled woof he leaped off the trail and climbed a steep hogsback. I hate to think what would have happened if the beast had been in an ugly mood as my rifle was on the saddle and I would not have had time to use it. People have often asked me why I did not carry a revolver for protection. A revolver is only effective at short range, and distance is what counts when you have an argument with a grizzly.

Bears are, I think, the most unpredictable animals under the sun. It is said that a she bear regardless of colour will always defend her young, but I have seen this theory disproved. Two of us were travelling in a pickup and surprised a bear and her cub. The mother bounded into the bush and left her offspring to its own resources. She did not even return to the whimpering of the poor little fellow as long as we stayed there.

Another case was totally different. Later, when I was Headquarters ranger, my son Jim and I were on our way to Winter Creek and I was driving the Forestry pickup truck. Rounding a sharp bend in the road I almost ran into a black bear and her two cubs. They started to run, but one cub darted into the bush and as soon as the mother saw this she turned around and made for us. Chattering and frothing at the mouth she travelled first on two legs and then four. I was ready to go in reverse, as tooting the horn only seemed to enrage her more. Luckily the wayward cub decided to come back to Mother, who in turn swatted him with her paw causing him to roll a couple of times. All three now made off and we saw no more of them.

Clarence Earl and I built a fly shelter for our horses at Rock Lake. The reason for this was that when the flies were troublesome, the horses would wander a long way up the Wildhay River to stand on a gravel bar. Here they would stand, one facing one way and one the other, their tails switching continuously. As the saying goes, "You scratch my back and I'll scratch yours." We figured that since we could always depend on finding them in the shed at Headquarters in fly season, they would do the same at Rock Lake, but it didn't work out that way. Days that we didn't travel, we would go out and catch them and tie them up in the shed during the day and turn them loose in the evening. However, they never did take kindly to the shed, so we gave up the idea in disgust.

That fall and winter, with the help of my wife, I built a log garage and woodshed and cut a year's supply of firewood. Things took a turn for the better with the discovery of oil in the area and a change of government, and in the fall I was taken back by the Forest Service on a permanent basis.

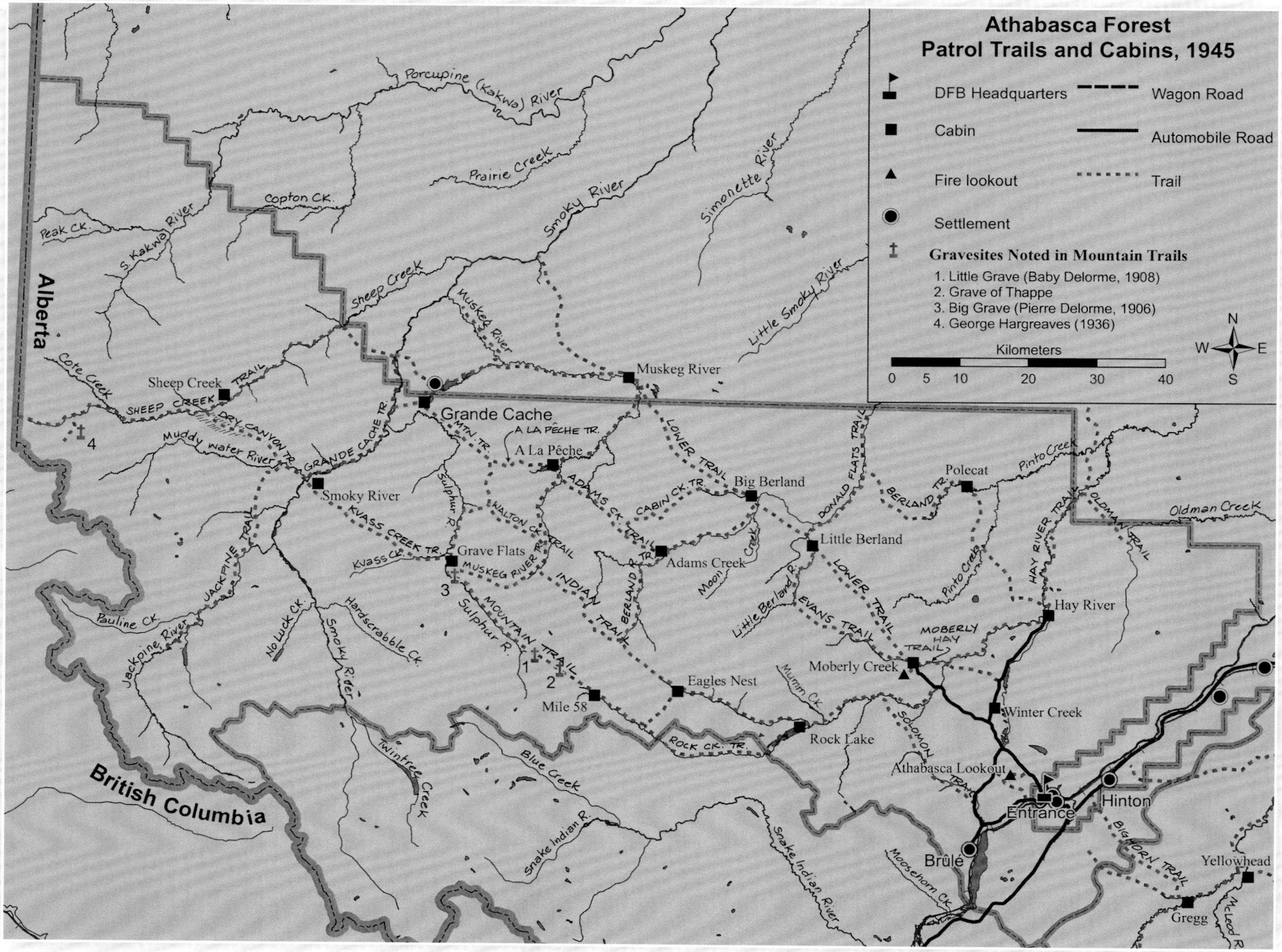

Athabasca Forest
Patrol Trails and Cabins, 1945
DFB Headquarters
Cabin
Fire lookout
Settlement
Wagon Road
Automobile Road
Trail
Gravesites Noted in Mountain Trails
1. Little Grave (Baby Delorme, 1908)
2. Grave of Thappe
3. Big Grave (Pierre Delorme, 1906)
4. George Hargreaves (1936)
Kilometers
0 5 10 20 30 40
N
W
E
S
Alberta
British Columbia
Grande Cache
Sheep Creek
Muskeg River
A La Pêche
Smoky River
Grave Flats
Big Berland
Adams Creek
Little Berland
Polecat
Hay River
Moberly Creek
Eagles Nest
Mile 58
Rock Lake
Winter Creek
Athabasca Lookout
Entrance
Hinton
Brûlé
Yellowhead
Gregg
Porcupine (Kakwa) River
Prairie Creek
Smoky River
Simonette River
Little Smoky River
Copton CK.
Peak CK.
S. Kakwa River
Sheep Creek
Cote Creek
Muskeg River
SHEEP CREEK TRAIL
DRY CANYON TR.
Muddy water River
GRANDE CACHE TR.
MTN. TR.
A LA PÊCHE TR.
LOWER TRAIL
DONALD FLATS TRAIL
Pinto Creek
BERLAND TR.
Oldman Creek
OLDMAN TRAIL
HAY RIVER TRAIL
CABIN CK. TR.
ADAMS CK. TRAIL
WALTON CK. TRAIL
Sulphur R.
KVASS CREEK TR.
Kvass CK.
MUSKEG RIVER TR.
INDIAN TRAIL
BERLAND R. TR.
Moon Creek
Little Berland R.
EVANS TRAIL
LOWER TRAIL
Pinto Creek
MOBERLY HAY TRAIL
Mumm CK.
SOLOMON
ATHABASCA TRAIL
MOUNTAIN TRAIL
Sulphur R.
JACKPINE TRAIL
Pauline CK.
Jackpine River
No Luck CK.
Smoky River
Hardscrabble CK.
ROCK CK. TR.
Twintree Creek
Blue Creek
Snake Indian R.
Snake Indian River
Moosehorn CK.
BIGHORN TRAIL
McLeod R.
1
2
3
4

Jack Glen (1891-1983), c. 1981. In retirement, Mr. Glen devoted himself to his memoirs, his family and his artwork.

JACK GLEN PHOTOS, ALBERTA FOREST HISTORY PHOTOGRAPHIC COLLECTION

Promotion and Last Patrol

Shortly after this, at the end of 1941, the Headquarters ranger L.L.Waikle, resigned, and since I was next in seniority I took over. It was wonderful to be home most of the time, though I missed the free-roving life that I had grown accustomed to over the years. With the help of some of the boys I built a new log horse barn, an equipment cache and implement shed, and moved the blacksmith shop to a new location.

One feature that I did not fancy about my new job was that when a railroad fire was reported, I had to use the Forestry speeder to get to it. As I have often remarked, I would sooner meet a grizzly bear on the trail than a locomotive on the track. However, as more roads were constructed, the need for the speeder grew less and less and finally it was disposed of in favour of a pickup truck, which did nothing to hurt my feelings.

I was kept busy with a multitude of duties. When I was not out on timber inspections, I would be collecting grazing dues or issuing angling and travelling permits. As our son grew older, my wife and I found that we had either to board him out in order to get his higher education, or to quit the job and move somewhere where he could obtain one. We chose the latter course, and in 1945 I left the Forest Service and moved to British Columbia.

I have been told that many changes have taken place with the discovery of oil in the Forest Reserve, but I like to remember it as it used to be. In fancy, I find myself quite often riding among those old familiar Mountain Trails.

Opposite page: Jack Glen's letter to former ranger John Currat mentions Percy Wroe who was a DFB forest ranger on the Athabasca Forest from 1925 to 1930. Judd Groat remarked that Wroe Creek was named after him, officially noted in 1945. Wroe Creek flows into Pinto Creek. From: Feddema-Jones 2007 and Karamitsanis 1991.

APPENDIX 1: A Letter from Jack Glen to John Currat, 25 September 1971

4817 Davis Road
Terrace, B.C.
Sep. 26 — 71

Dear Mr. Currat.

I wish to thank you for your most interesting letter and enclosure. Unfortunately my memory is failing me and as regards the information you requested I'm afraid I won't be of much help but I will do my best. When I first joined the Forest Service early in 1920 it was a Dominion show and in its infancy, also the Athabaska Forest Reserve had practically ceased to exist during the First World War, so it was pretty much like starting from scratch. I was posted to the Mountain District, which, as you know did not have telephone service. Mr Badgley was Supervisor and he briefed me on my new duties, particulary on the hazards of fording swollen rivers, Grizzly encounters and Bull moose during rutting season. The salary at that time was a mere pittance and work was plentiful with fairly good wages so the Service suffered. It was the practice then to have two men travelling together, a ranger and assistant. Mr. Badgley didn't like the idea of sending me (a greenhorn) out alone but he promised that he would hire the first available man as my assistant and he was as good as his word. Rock Lake was the only cabin in the entire district and it had been fixed up the previous year but the ranger had quit. I was alone for about two months so I spent the time in getting to know the district and mountain climbing. Finally I was joined by an old trapper who brought me a letter of introduction and requesting us to complete a cabin at the foot of Eagle Nest Pass. It lacked three rounds and had a terrible odor of pack rat. The logs were unpeeled so I attended to that while my partner cut and skidded poles for the roof. The Provincial Game Guardian wished to observe conditions in the Provincial Reserve and as usual couldn't find a helper. Mr. Badgley asked if I would care to accompany him so needless to say I jumped at the chance. At the end of that season I quit the Service, but returned two years later to resume where I had left off. You mentioned reading my story so now I will just list the various projects that I completed. As I mentioned Rock Lake Cabin existed as did the Mountain and Lower Trails which as far as I am aware were built by a man by the name of Bowman. I completed Eagle Gates with assistance of Blackey Lanoil. I completed Eagle Nest Cabin, built Summit Cabin and later the Smoky River Cabin. Ranger Holm built Grave Flats Cabin and possibly Grand Cache Cabin. A La Peche Lake Cabin, New Winter Creek and Moberly L.O. Charlie Milindorf built Little Berland Cabin. As far as I know the telephone line from Entrance to the Muskeg River Cabin was let by contract. Percy Uroe and Bill Douglas built the Adams Ck. Cabin and the Adams Ck. Trail was cut by myself and trail crew, together with Muskeg River Trail, Kvass Ck. Trail, Sheep Ck. Trail, Sulphur River Trail and Berland River Trail to the 15th Base. I located and hauled all buil'ng material for the Athabaska L.O. during Winter and with my crew in the Spring we cut and built and installed the telephone in the building which was erected by a carpenter named McIvor. We also cut the line and strung the wire to the Moberly L.O. Tom Coggins hauled the material for the Berland L.O. I forgot to mention that I located and cut a trail from Hay River Cabin to Donald's Flats on the Berland River. That was during the hungry thirties and no more trail crews were hired but a year or so later I was instructing a class on Forestry and we extended the Telephone line from Winter Ck to Hay River Cabin.

I am enclosing a few snaps that might be of interest to you I can't locate the negatives of the small ones but you may keep the two large ones.

I have been wondering if Mr. Murphy would care to have my story published in book form. It could possibly be rewritten and more detail added. If this could be acheived it would be most gratifying as I have been deluged with fan mail and nearly all wishing to know where the book could be bought.

I am sorry that I can't remember the Telepone rings except that Winter Ck. was 1—#1— but I could be wrong. Many of the old boys have passed on and I left Entrance 26 years ago with most comunication being transmitted by radio telephone. Have you tried Gordie Watt, Bert Prowse, (if still alive.) Ted Hummer might know. He lives in Kelowna.

Thanks again

Sincerely Jack Glen.

APPENDIX 2: Jack Glen, Forest Ranger,* History of Employment, 1920 to 1945

Until the end of the 1952/53 fiscal year, government financial departments listed the names of employees, their positions and salaries or wages. This table summarizes those for Jack Glen.

Year	Position	Forest	Rate	Amount (for period of service)	# Months
1920/21	Forest Ranger**	Athabaska	$100 per month	4 May to 31 October + bonus $79.69	5
1921/22	Not listed				0
1922/23	Forest Ranger	Athabaska	$100 per month	5 October to 31 October	1
1923/24	Forest Ranger	Athabaska	$1200 per annum	1 May to 31 March***	11
1924/25	Forest Ranger	Athabaska	$1200 per annum	$1200 + $96 forage	12
1925/26	Forest Ranger	Athabaska	$1320 per annum	$1320	12
1926/27	Forest Ranger	Athabaska	$1440 per annum	$1440	12
1927/28	Forest Ranger	Athabaska	$1680 per annum	$1680	12
1928/29	Forest Ranger	Athabaska	$1800 per annum	$1800	12
1929/30	Forest Ranger	Athabaska	$1800 per annum	$1800	12
1930/31	Forest Ranger, DFB	Athabaska	$1800 per annum	April to November 1930	8
1930/31****	Forestry Assistant, AFS	Athabaska	$1800 per annum	December 1930 to March 1931	4
1931/32	Forestry Assistant	Athabaska	$1900 per annum	$1899.60	12
1932/33	Forest Ranger	Brazeau-Athabasca	$150 per month	$1200.00	8.0
1933/34	Forest Ranger	Brazeau-Athabasca	$150 per month	$1016.00	6.8
1934/35	Forest Ranger	Brazeau-Athabasca	$150 per month	$900.00	6.0
1035/36	Forest Ranger	Brazeau-Athabasca	$150 per month	$975.81	6.5
1936/37	Forest Ranger	Brazeau-Athabasca	$150 per month	$1016.13	6.8
1937/38	Forest Ranger	Brazeau-Athabasca	$150 per month	$1118.61	7.5
1938/39	Forest Ranger	Brazeau-Athabasca	$150 per month	$1040.32	7.0
1939/40	Forest Ranger	Brazeau-Athabasca	$150 per month	$1198.34	8.0
1940/41	Forest Ranger	Brazeau-Athabasca	$150 per month	$1090.55	7.3
1941/42	Forest Ranger	Brazeau-Athabasca	$150 per month	$1775.32	12
1942/43	Forest Ranger	Brazeau-Athabasca	$125 per month	$1620.00	12
1943/44	Forest Ranger	Brazeau-Athabasca	$1680 per annum	$1815.00	12
1944/45	Forest Ranger	Brazeau-Athabasca	$1680 per annum	$1796.79	12
1945/46	Not listed				

* Records extracted from Public Accounts database by P.J. Murphy, 2004-09-11.

** He was first listed in the 1920/21 fiscal year, confirming the date in his memoirs. Listed as J. Glenn. He was not listed for 1919/20.

*** Record in my set is missing – have estimated based on the two adjacent years [pjm].

**** This was the year of the Transfer of Resources – DFB evidently paid to end November 1930.

Comments:

Glen started with the Dominion Forestry Branch in early May 1920, as he described in his memoirs, ending his work that first year at what may have been the end of the fire season on 31 October 1920. When he mentioned that he had quit DFB for two years, this seems to have been the time he did it. His name is not listed for 1921/22. He evidently came back for the month of October 1922, then started again on 1 May 1923. We can only conjecture where he spent the winter from November 1922 to April 1923. He then continued right through to the end of March 1945, which is later than indicated in his memoirs. Perhaps he ended his memoirs in 1942 as the original sub-title indicated, but his employment record continues for another two full summer seasons.

His record also shows that the Transfer in 1930 went seamlessly for him, with no lost time in the process. DFB evidently paid until the end of November 1930 and AFS started him in December. However, the effect of the Great Depression is clearly evident in the summary. In the fiscal year 1932/33 annual salary was converted to monthly wages, the monthly rate was reduced and the period of employment clearly became seasonal, as low as 6 months in 1934. In 1942 wages were cut again to $125 per month. In 1924 there appears to have been a forage allowance, presumably to help with his horses.

Sources:

1920–1930: Annual Reports of the Auditor General for the Department of the Interior, Forestry Branch, Canada.

1930–1945: Annual Reports of the Public Accounts, Department of Treasury, for the Department of Lands and Mines, Alberta Forest Service, Alberta.

APPENDIX 3: Local friends and neighbours attending Jack Glen's farewell party, Old Entrance, 28 February 1945

A number of people with strong ties to the early history of Hinton and Entrance gathered on 28 February 1945 to give Jack Glen and his family a good send-off as he had resigned from the Forest Service and was moving to British Columbia. The following names, listed in Mr. Glen's handwriting, were provided by Jim Glen from his personal files. The editors have added further details in brackets where full names could be determined or seemed most probable from names in the *History of Hinton* (Hazel Hart, 1975) or the AFS employment records.

Mr. & Mrs. R(ufe) Neighbor
Bert Davey
Wilbert Magee
Mr. & Mrs. (George) Munro
Mr. & Mrs. R(aymond) Fuller
Mr. & Mrs. H(arry) Hammer
Mr. & Mrs. P. Dumenko
Mr. & Mrs. C(harlie) Matheson
Mr. Cyrus Mason
Mr. H(arold) Lake
Mr. H(arry) Davison
Miss M(ildred) Woodley
Mr. & Mrs. C(harlie) Bowlen
Mr. & Mrs. R(uben) Kilba
Mrs. P.H. Kreye
Mr. R. Rhein
Mr. Ludwig Hoff
Mr. & Mrs. R(oy) Woodley
Mr. & Mrs. M(ark) Truxler
Mr. & Mrs. C(harles) Chapman
Mr. & Mrs. H(arry) Ennis
Mr. & Mrs. C(arl) Larson
Mr. A.H. (Bert) Prowse
Mr. Lonzo Langill (*sic*) – (Len "Blackie" Langelle)
Mrs. H(ans) Peterson – (formerly Eileen Woodley)
Mr. & Mrs. R(alph) Cropley
Mr. & Mrs. J(ack) MacLean
Mr. Shand Harvey

APPENDIX 4: Rangers and Service Years, Athabasca Forest, 1914 to 1945

Name Surname	Name Initials	Name Given	Position	First Year	Last Year	# Months	Notes
Fiddis	J.W.		Fire Ranger	1912	1912	5	Forest Fire Ranger at Hinton
Harvey	J.S.	James Shand	Forest Ranger	1912	1917	16	James Shand Harvey was a ranger in 1916, subject of McGregor's Pack Saddles to the Tête Jaune Cache
Hutchison	J.A.	John	Fire Ranger	1912	1912	8	Transferred, later became Assistant Director of Forestry
MacFayden	C.F.		Forest Supervisor	1912	1917	34	MacFayden was Supervisor 1916 –1917
Scidmore	E.C.		Fire Ranger	1912	1912	4	
Bessette	O.U.		Seasonal	1913	1913	4	Time estimated – appears seasonal
Clark	S.H. (Stan)	Stanley H.	Forest Supervisor	1913	1919	75	U. of Toronto forester, Class of 1913, first designated supervisor of Athabasca Forest, including 2 years WW1 and Brazeau to July 1919[75]
Colwell	G.E.		Forest Ranger	1913	1913	4	
Hannum	R.		Forest Ranger	1913	1913	5	
Harrison	E.A. (Ernie)	Ernest Alexandar	Forest Ranger	1913	1939	163	Harrison worked 1913, 1917 then again in 1922-1939
Langill	L.	Len	Forest Ranger	1913	1916	10	1913 listed as Langhill, L. – probably the same as Langill, L.
Smith	J.M.		Forest Ranger	1913	1914	23	
Anderson	J.		Forest Ranger	1914	1915	19	
Bothwell	G.E.		Forest Assistant	1914	1916	30	U. of Toronto forester, Class of 1913
Gaetz	J.B.	(Bowman)	Forest Ranger	1914	1916	31	
Harron	J.J.		Forest Ranger	1914	1918	55	
Johnston	J.J.		Forest Ranger	1914	1914	2	
Mason	C.F.		Forest Ranger	1914	1915	10	
Leatham	J.	Jefferson	Forest Ranger	1915	1918	14	
Millar	H.S.		Clerk	1915	1916	12	
Monaghan	T.	Tom	Forest Ranger	1915	1921	10	Guide, outfitter and trapper, opened a store at Entrance
Olson	B.	Benjamin	Forest Ranger	1915	1917	23	
Bellack	K		Forest Ranger	1916	1916	4	
Gregg	J.		Forest Ranger	1916	1916	4	
Thompson	W.E.		Forest Ranger	1916	1917	15	
Warner	J.W.		Forest Ranger	1916	1916	4	
Welch	T.H.		Forest Ranger	1916	1917	6	
Badgley	W.W. (Ward)	William Ward	Forest Supervisor	1917	1922	56	Started April 1914 at Slave Lake, Athabasca 1 August 1917 to 31 March 1922
Hedberg	J.	John	Forest Ranger	1917	1918	10	
Hindmarsh	J.	James	Forest Ranger	1917	1917	3	
Raymond	F.		Forest Ranger	1917	1917	2	
Severson	O.		Forest Ranger	1917	1917	11	
Hawkey	R.E.		Forest Ranger	1918	1921	38	

Name Surname	Name Initials	Name Given	Position	First Year	Last Year	# Months	Notes
Holm	L.	Louie	Forest Ranger	1918	1929	132	
Lindgren	J.		Forest Ranger	1918	1918	5	
Milindorf	C.F. (Charlie)	Charles	Forest Ranger	1918	1921	23	
Molstrom	P.		Forest Ranger	1918	1919	17	
Whimster	H.M.		Forest Ranger	1918	1918	8	
Adams	D.		Forest Ranger	1919	1920	11	
Baillie	O.C.		Forest Ranger	1919	1920	10	
Cooley	I.		Forest Ranger	1919	1919	1	
Fitzgerald	F.		Forest Ranger	1919	1919	5	
Dickey	G.		Teamster	1920	1920	4	
Glen	J. (Jack)	John	Forest Ranger	1920	1945	227	Jack Glen, author of Mountain Trails. Longest-serving ranger in Athabasca Forest as of spring 1945
Hammerirck	S.M.		Forest Ranger	1920	1920	5	
Horfst	J.D.		Forest Ranger	1920	1920	5	
Salter	W.		Forest Ranger	1920	1923	30	
Marquette	R.I.		Forest Ranger	1921	1921	7	
Mills	J.H.	Jim	Forest Ranger	1921	1922	18	
Plante	F.	(Felix)	Forest Ranger	1921	1921	1	May have been Felix Plante
Wheeler	J.E.		Forest Ranger	1921	1921	7	
Burrows	T.C.	Tom	Forest Supervisor	1922	1933	126	
Cook	J.E.		Forest Ranger	1922	1922	5	
Galbraith	A.		Forest Ranger	1922	1922	4	
Irwin	T.N.		Forest Ranger	1922	1922	10	
McCauley	G.		Forest Ranger	1922	1922	1	
Wells	W.C.		Forest Ranger	1922	1922	4	
Douglas	W.W. (Bill)	William	Forest Ranger	1923	1929	69	
Holgate	R.W. (Bob)	Robert William	Forest Ranger	1923	1937	90	
Ruzicka	A.R.G.		Forest Ranger	1923	1923	6	
Weirzba	William		Forest Ranger	1923	1923	6	
Griggs	R.A.		Forest Ranger	1924	1924	5	
Martelle	J.		Forest Ranger	1924	1925	8	
Reader	L.H.		Forest Ranger	1924	1925	12	
Wroe	P. (Percy)	Percy	Forest Ranger	1925	1930	42	
Campbell	R.J.		Forest Ranger	1926	1927	10	
Hammer	A.H. (Harry)	Arius Harry	Forest Ranger	1926	1945	111	Brother of Ted Hammer, was also a lookoutman, transferred to Hinton 1941 – 1944
Huestis	E.S.	Eric	Forest Engineer	1926	1928	26	Transferred – later became Director of Forestry, Deputy Minister of Lands and Forests

Name Surname	Name Initials	Name Given	Position	First Year	Last Year	# Months	Notes
Jackson	F.H.R.		Forest Ranger	1928	1930	27	
Smith	W. (Bill)	William	Forest Ranger	1928	1945	118	
Coggins	T.F. (Tom)	Thomas Frank	Forest Ranger	1930	1945	129	
Hammer	T.R. (Ted)	Theodore R.	Forest Ranger	1930	1940	75	Transferred to Grande Prairie, head of forest protection c. 1955
Reimer	A.	(Abe)	Forest Ranger	1931	1932	12	Probably Abe Reimer, guide and outfitter
Teitge	F.W. (Fred)	Frederick William	Forest Ranger	1931	1945	115	
Edgar	F.C. (Fred)		Forest Supervisor	1933	1935	36	
Waikle	L.L.		Forest Ranger	1933	1942	50	Moved within Brazeau-Athabasca, left 1942
Smith	A.G.		Superintendent	1935	1938	40	Supervisor of Athabasca from June 1935
Donnelly	P. (Pat)	Patrick	Forest Ranger	1937	1939	18	Later became radio operator, then aircraft dispatcher with AFS
McCardell	W.H.	Bill	Forest Ranger	1937	1939	18	His father was co-discoverer of Cave & Basin in Banff
Frew	I. (Irv)	Irvine Bell	Forest Ranger	1939	1940	4	Started in Crow-Bow in 1934, transferred in and out of Athabasca in 1939 and 1940
Huestis	E.S.	Eric	Supervisor	1939	1940	13	Later director of forestry, Deputy Minister of Lands & Forests
Bradshaw	F.C. (Frank)		Forest Ranger	1940	1940	6	
Larson	C.	Carl	Forest Ranger	1940	1944	24	Transferred within AFS, died in a river drowning in 1950s
Barrow	E.A.		Forest Ranger	1941	1942	12	
Buck	W.F.D.	Donald	Superintendent	1941	1945	63	Forest Superintendent at Edson from January 1941 to 1956
Burleigh	J.H. (Red)	James Harold	Forest Ranger	1941	1942	20	Transferred within AFS
Earl	C.E. (Clarence)	Clarence Ethbert	Forest Ranger	1941	1941	8	Transferred within AFS, on 1942 Brazeau-Athabasca list
Falk	C.		Forest Ranger	1941	1942	12	
Chapman	C. (Charlie)	Charles	Forest Ranger	1942	1945	36	
Prowse	A.H. (Bert)	Albert Henry	Forest Ranger	1942	1945	60	
Sherman	A.		Forest Ranger	1942	1943	21	
Watt	A.G. (Gordon)	Andrew Gordon	Forest Ranger	1942	1945	48	Left AFS in 1950 to run his store at Entrance, formerly Monaghan's
Lewis	R.T.	Bob	Forest Ranger	1943	1944	18	
Haight	A.R.		Forest Ranger	1945	1946	12	
Richardson	W.L. (Walt)	Walter	Forest Ranger	1945	1946	12	Replaced Jack Glen. Later Chief Scaler for NWPP
Wright	R.G.		Forest Ranger	1945	1946	12	

INDEX

A

À La Pêche Creek, 156, 158
À La Pêche Lake, 65, 156
À La Pêche Trail, 155, 158
Abbotsford, B.C., 139
Adams Creek, 128
Adams Creek Trail, 128, 146, 147
Adams, D., 292
Alberta Forest Ranger, **12**
Alberta Forest Service, 226, 230, 266, 270, **282**, 290, 293
Alberta Government Telephones, 10
Alberta Sustainable Resource Development, 98
Anderson, J., 291
Association Saddle, 96, 98
Atfield, Paul, 107
Athabasca Bridge, 17, **152, 153**
Athabasca Forest (Reserve), **11,** 13,14,**17,** 18, **19, 22,** 39,49, **52,** 63, **76, 80, 81,** 84, **85,** 86, 89, 93, **94,** 96, **106**, 107, 119, 126, 142, 146, 147, 153, 165, 178, 192, 210, 226, 238, 248, 253, 267, 270, 272, 278, 282, 284, 291, 292,
Athabasca Forest boundary, 142, 146
Athabasca Lookout, 121, **122,** 123, 124, 126, 127, 169, 244
Athabasca Ranch, 106, 134
Athabasca River, 29, 73, **102, 103, 115**, 134, 135, 166, 185, 186, 187, 196, 264, **271**
Athabasca Valley, 127
Atlantic Ocean, 16

B

Babiche, 70
Badgley, Edna, 139
Badgley, William Ward, 13, 14, 19, 20, 21, 22, 27, 29, 47, 49, 63, 71, 78, 79, 291
Bailey, "Terk", **266**
Baillie, O.C., 292
Banff, Alberta, 274, 293
Baptiste River, 219
Barrie, Ontario, 14
Barrow, E.A., 293
Baseline, 14th, 53; 15th, 84, 102, 103, 282
Battleford, Saskatchewan, 15
Beaudry, Father, 116
Bellack, K., 291
Berland River, 18, **100,** 101, **142, 144, 145, 147, 169** 219, 282
Berland River District, 79, 210, 211, 264
Berland River Flats, **197**
Berland River Trail, **140**, 282
Bessette, O.U., 291
Bickerdike, 89
Big Berland country, 128, 253
Big Berland River, 36, 82, 84, 142, 146
Big Berland River Cabin, 64
Big Grave Flats, **38, 39, 41,82,** 104, 148, 151, 198, 200, 248, 249202, 244, 272
Big Grave Flats Cabin, 233, 243
Big Smoky district, 72
Big Smoky River, 71, 104, 105, 200, 208, 212, 225, 230, 233, 238
Bill, Buffalo, 58
Bingen, Bishop of, 148
Biznet, Joe, 84, 85
Black Cat (Brûlé Cat), 16, 17
Black Cat Ranch, 114
Blackfoot Indians, 15
Blackie *(see Daw, Langelle)*, 159
Blue Diamond Mining Company, 18, **47**
Boone, Daniel, 58
Bothwell, George E., 106, 291
Boulder Creek, 174, 202, 231
Boulder Creek Fire, 230
Boulder Creek Trail, 174
Boundary survey camp, 1911, **102-103**
Bow River Forest, 89, 99, 100, 270, 274
Bowlen, Charlie & Mrs., 290
Bradshaw, Frank C., 293
Brazeau Forest, **64, 88, 113, 115**, 123, 153, **235,**
Brazeau River, 49, 291
Brazeau-Athabasca Forest, 266, 293
Brewster, Fred, 253
Brewster, Jack, **42, 43.** 114, 253
Brewster, Major Fred, 114, 244, 279
Brewster's Wall, 114
British Columbia, 151, 181
British North America Act, 1867, 26
Brule Lake, 18, **135**
Brule, Alberta, 16, 17, **47**, 48, 55, 56, 138, 210, 275, 278
Buck, W.F.D. (Donald), 293
Burleigh, J.H. (Red) James Harold, 293
Burns Lake, B.C., 86
Burns, Robbie, 22
Burrows, Mrs. T., 96
Burrows, Tom C., 79, **80-81**, 84, 86, 96, 104, 120, 121, 122, 124, 153, 166, 205, 218, 225, 292
Byng of Vimy, Baron (Gov. General), 119

C

Cabin Creek, 274
Calgary Tigers, 270
Calgary, Alberta, 22, 252, 272, 238
Calliou clan, 157
Calmar, Alberta, **78**, 82
Campbell, Kenneth, 104, 105, 108
Campbell, R.H., **54**
Campbell, R.J., 293
Canadian Forest Service, 107
Canadian National Railway, 14, 16, 134, 135, 139, 166
Canadian Northern Railroad, **47, 168, 271**
Carcajou, 175
Carman, Bob, 107
Carson Creek, 114
Caruso, 90
Casket Creek, 181
Casket Mountain, **182-183**
Cathedral Mountain, Eagles Nest Pass, **28**
Cave and Basin, 293
Cayuse Indian Pony, 22
Chapman, C. (Charlie) Charles, 293
Chapman, Charles & Mrs., 290
Chapman, Eileen, **276, 277**
Chilkoot Pass Slide, 16
Chinese workers 92-93
Chinook, 191
Chip Lake, 139
Christie, John and Rita, **168**
Churchill, Sir Winston, 134
Circle "M" Ranch, **119**
Circus, Barnum and Bailey, 126
Clara, Indian girl, 138
Clark, Constable Archie, 227, 233, 234, 267, 270
Clark, Stanley H. (Stan), **85,** 105, 106, **107, 150, 169, 176,** 134, 212, 241, 250, 265, 267, 273, 278, 291
Clark's Crossing, 106, **170, 171, 175, 212, 234**
Clearwater Forest, **121**, 266
Coal Branch, Alberta, 227
Coal Creek logging camp, **89**
Coalspur, Alberta, **123**, 225
Coffin Creek *(see also Casket Creek)*, 181
Coffin Mountain *(see also Casket Mountain)*, 181
Coggins, Thomas Frank (Tom), 210, 212, 213, 217, 218, 293
Coliseum Lookout, **121**, 194
Cook, J.E., **80-81**, 292
Cook, Thomas, 147
Cooley, I., 292
Corral Creek, 217
Corrigan, Douglas "Wrong Way", 212-213
Cowlick Creek, 68, 158
Colwell, G.E., 291
Coyote Cliff *(see also Faulk, Cliff)*, 202
Cree Indians, 15, 33, 53, **54**, 156, 175
Crockett, Davy, 58
Cropley, Ralph. & Mrs., 290

Crosby, Bing, **160, 164, 165**
Crossley, Des, 107
Crow-Bow (Forest), 293
Crowsnest Forest, 233, 266
Crowsnest Pass, Alberta, 252
Currat, John, 284, 285

D
Davey, Bert, 106, **204**, 222, 224, 290,
Davison, Anne, 134
Davison, Frankie, 134
Davison, Harry S., 16, 106, 134, 152, 290
Daw, Blackie, 74
Delorme, Baby, 33
Delorme, Marie, 157
Delorme, Pierre, **82**
Depression, The Great, 238, 266, 289
Devona Warden Station, **55**
Diamond Hitch, **193**
Dickey, G., 292
Dominion Forestry Branch, **12, 15, 17, 19, 22,** 49, **54, 72,** 73, **80**, 96, **104**, 106, 123, 126, 135, 211, 226, 272, 284, 289
Dominion government, 31, 56
Dominion Parks Board, 39
Donald Flats, 84, 101, **142, 143, 144,145, 178**,198, 282
Donald Flats Trail, 282
Donnelly, Patrick (Pat), 293
Douglas, William W. (Bill), 89, 93, 104, 123, 128, 153, 204, 292
Drinnan Mine, 278
Dry Canyon, **109**, 174, 213
Dry Canyon Trail, 282
Dumenko, Mr. & Mrs. P., 290
Dutch oven, 48, 56
Dyke railway station (GTPR), 139
Dynamite (horse), 193

E
E.F. Woodley General Merchandise, 139
Eagles Nest, 34, 47, 72, 147, 194, 225, 259, 272
Eagles Nest Cabin, 27, 33, 36, 59, 196, **201**, 238, 244, 253, 273
Eagles Nest Pass, 27, 200, 203, 257, **258**
Earl, Clarence Ethbert , 278, 279, 293, 282
Earnshaw, Tony, 233
East Park Gate, (Jasper), 274 (278?)
Edgar, Frederick George (Fred), 153, 225, 227, 293
Edmonton, Alberta, 10, 13, 47, 78, 86, 89, 91, 93, 133, 143, 169, 200, 204, 221, 226, 252, 272, 278
Edson Forest, 293
Edson, Alberta, 17, 224, 233, 272, 274
Edwards, Henrietta Muir, 26
Embarras River, 89
Empson, Don, 127, 143
England, 153
Ennis, Harry. & Mrs., 290
Entrance District, 199
Entrance Headquarters, 33, **63, 117,** 133, 230, **275**
Entrance Post Office, 139
Entrance Ranch, 16, 134
Entrance station, 139, **170**
Entrance, Alberta, **15, 16, 17,** 22, 27, 29, 33, 39, 47, 53, 59, 63, 65, **66, 67,** 73, 75, 79, 83, 86, 91, 93, 96, 101, 106, 113, **115**, 116, 117, 118, 120, 126, 133, **134**, 135, 138, 139, 143, 147, 150, 151,152, 155, 166, 169, 172, 175, 185, 190, 193, 196, 198, 132, 200, 202, 203, 204, 205, 210, 211, 217, 218, 222, **223,** 224, 225, 227, 230, 231, 233, 234, 235, 238, 244, 252, 253, 259, 260, 262, 264, 267, 270, **271**, 273, 274, 275, **276, 277**, 278, 279, 281,
Entwistle, Alberta, 86
Ertman, Les, 98
Eton, U.K., 53
Evansburg, Alberta, 86

F
Falk, C. *(see Faulk, C.)*, 293
Famm Creek, 112, 174
Famous Five, 26
Faulk Creek, 202
Faulk, Cliff, trapper, 175, 202, **203,** 250, 273
Fay Expedition, 1914, **82, 148, 191, 216**
Felix Creek, 253
Fiddis, J.W., 291
Findlay, Jacco, 69, 84
Findlay, Madeleine, 68
Fire Notice. **54, 237**
First World War, 10, 106, 291
Fish Creek Flats, 138
Fish Lake Flats, 138
Fitzgerald, F., 292
Foothills Model Forest, 98
Forest Rangers, **66, 67, 80, 81**
Forest Reserve, 47, 48, 58, 63, 71, 78, 253, 257, 284
Forest Service, *(see also Dominion Forestry Branch, Alberta Forest Service)*62, 96, 243, 235, 236, **242,** 250, 272, 282, 284, 290
Forestry Training School, 226
Fort McMurray, Alberta, 267
Fort McPherson, NWT, 204
Fort Saskatchewan, Alberta, 185
Fraser River, 91
Freedom, Alberta, 204
Frew, Irving Bell (Irv) , 270, 271, 272, 293
Fuller, Raymond. & Mrs., 290

G
Gaetz, J.B. (Bowman), 291
Galbraith, A., 292
Galbraith, Scotty, 92
Georgia, (USA), 17
Germany, 265
Ghost River Fire, 214
Gilliat, Neil, 73, 178, 202, 227
Glacier Creek, 192
Gladstone, Nellie, 49
Glen, Jim, 224, 275, 279, 282, 276, 277, 280, 281, **284-5**
Glen, John (Jack), **32, 78, 97,** 123, **130, 131,** 172, 199, 200, 208, 215, 230, 235, **263, 264, 268, 269, 280, 281,** 282, **284,** 290, 291, 284, 285, 286, 287, 290, 291, 292
Glen, John (Jack) – first foot patrol – map, 37
Glen, John (Jack), history of employment, 288
Goat Camp, 105, 231
Goldsmith, Oliver, 224
Grand Trunk Pacific Railroad, **16, 86, 134, 135**, 271
Grande Cache Cabin, 129, 132, 191, 213
Grande Cache, Alberta, 38, 39, 40, 49, **68, 71,** 83, 157, 172, 178, 184, 185, 217, 230, 238, 239, 240, 247, 272
Grande Prairie, Alberta, 226, 234, 247, 293
Grave Flats, 148, 149, 250
Grave Flats Cabin, 203, 217, 227, 232, 238, 250, 265, 272, 279
Grave Flats Lookout, 126
Greenwood, Ranger, 235
Gregg, J., 291
Grey, Joseph, 157
Grey, Pierre (or Peter), 156, 157
Griggs, R.A., 292
Groat, Judd, 284
Groat, Thomas, 49

H
Hadley, Mr., 86
Haight, A.R., 293
Hall, Mr., **180**
Hammer, A.H. (Harry), 226, 272, 290, 293
Hammer, Theodore R. (Ted), 15, 224, 226, 227, 230, 233, 238, 239, 240, 243, 244, 246, 251, 252, 253, 256, 259, 267, **268, 269, 270, 271**, 272, **275**, 293
Hammerirck, S.M., 292
Hanington, Bill, 137, 274
Hanington, Jacqueline (Jackie) (*nee* Truxler), 273, 274
Hannum, R., 291
Hardscrabble Creek, 207
Hardscrabble Pass, 34, **205-207**
Hargreaves brothers (Frank, Dick, Roy), 181

Hargreaves, George, **180, 181**
Hargreaves, Roy, **190**
Harold Lake, 202
Harragin, Agnes, **119,** 274
Harragin, Mona, **119,** 274
Harrison, Blanche, **168, 222, 223**
Harrison, Ernest Alexander (Ernie), **80, 81,** 96, 104, 105, 109, 111, 115, 116, 117, 124, 133, 136, **159, 162,163, 166, 172,** 218, 236, 241, 243, 252, 291
Harrison, Ranger Wally, **228, 229**
Harron, J.J., 291
Hart, Hazel, 53
Harvey, James Shand, 38, **73,** 53, 233, 234, 290, 291
Hawaiian Islands, 253
Hawkey, R.E., 63, 72, 292
Hedberg, John, 292
Henderson, Cliff, 98
Hendrickson Creek, 126
Hendrickson, Fred, **122,** 124, **125**, 126, **127,** 172, 191, 244, 264, 275
Herron, John, 49
Highway 40, 126, 135
Highwood River, **89**
Highwood River Camp, 266
Highwood-Pekisko Creek road, **100**
Hindmarsh, James, 292
Hinton Golf Course, 226
Hinton Hotel, 227
Hinton Ranger Station, **107,** 203
Hinton, Alberta, 16, 38, **65,** 107, 123, 134, 135, 139, 183, 226, 227, 274, 278, 290, 291
Hoff Ridge, 203
Hoff, Ludwig, 203, 290
Holgate, Robert William (Bob), 236, 238, 239, 243, 246, 292
Hollywood, 49, 165
Holm, Louie, **18, 22, 29,** 79, 155, 198, 200, 204, 292
Holmes, Mike and Mrs., 63
Hoppe Coal Claims, 178
Hoppe, Dr., 178
Horfst, J.D., 292
Huestis, E.S. (Eric), 293
Hughes, Stan, **266**
Hutchison, J.A. (John), 291

I
Idaho, USA, 106, 252
Iowa, USA, 62
Ireland, 213
Irish Sweepstake, 202
Irwin, T.N., **80, 81,** 292
Isaac Walton League, 126
Isle Lake, 157

J
Jack Pine district, 72
Jackson, F.H.R., 293
Jarvis Lake, 138
Jasper Highway, 86, 134, 152, 166, 200
Jasper Lake, **41**
Jasper Métis, 38
Jasper National Park, 33, 38, 39, 47, 48, 49, **52,** 55, 68, 71, 72, 75, 106, 160, 165, 197, 200, 204, 221, 253, 256, 257, 259, 266
Jasper Park Lodge, 165
Jasper, Alberta, 17, 114, 118, 119, 120, 157, 165, 181219, 221, 224, 253, 270, 271, 273, 274
Joachim, Frank, 165
Johnnie (Indian), 116, 124
Johnson, Albert "The Mad Trapper", 204
Johnson, E. B., 187
Johnston, J.J., 291

K
Kilba, Ruben. & Mrs., 290
King Edward Hotel, 13
Kinney, Rev. George, 119
Kitchen, Stan, **164, 165**
Klondike Gold Rush, 49
Kreye, Mrs. P.H., 290
Kvass Creek, **58**
Kvass Creek Trail, 146, 148, 208
Kvass District, 149
Kvass Summit, **104,** 148, 238, 243, 250

L
Lac Ste. Anne, 86, 138, 157, 253
Lady Godiva, 247
Lake Linderman, **199**
Lake, Harold, 175, 190, 264, 290
Lama, Les, **234**
Langelle, Len "Blackie", **73,** 75, **271,** 291
Langill, Len *see Langelle, Len "Blackie"*)73, 291
Langill, Lonzo (*see Langelle, Len "Blackie"*), 290
Larson, Carl, 290, 293
Lassie, 85
Leatham, Jefferson, 291
Leckie logger shoes, 159
Lewis, R.T. (Bob), 293
Lindgren, J., 292
Little Berland Cabin, 84, 142
Little Berland River, 33, 65, 84, **140,** 142, 198
Little Bo Peep, 85
Little Fort, B.C., 261
Little Grave District, 148
Little Grave Flats, 33, **34,** 202, 244, 279
Little Heaven Summit, 56, 58
Little Shushwap River , 93
Little Smoky River, 30
Loch Ness Monster, 267
London, UK, 267
Long Beach, California, 213
Longfellow, Henry Wadsworth, 134
Lovett Lookout, 126
Lower Trail, 17, 18, 19, 39, **64,** 65, **71, 82, 128, 132, 210**
Loyal Order of Dismounted Forest Rangers, 22
Luger, Carl, 139
Luger, Mary, 14, 134, 139
Luger, Mildred (Woodley), 139

M
MacDonald, Ranger, 123
MacFayden, C.F., 291
MacLean, Jack & Mrs., 290
Magee, Wilbert (Billy), 290
Maligne Lake, 114
Marie (teacher), 275
Marquette, R.I., USA, 292
Martelle, J., 292
Mary Reimer Park, 183
Mason, C.F., 291
Mason, Cyrus, 290
Massey Sawyer steam engine, 14
Matheson, Charlie, 112
Matheson, Charlie. & Mrs. (Mona), 290
Matheson, Mona (Harragin), 112, **119**
Mauritius, 53
Mayhoe, Johnny, 191
McCardell, W.H. (Bill), 293
McCauley, G., 292
McClung, Nellie (Janey Canuck), 23, 26
McDonald, Donald. 84, 178
McDougall, Mr. (sawmill owner), 93
McIvor, Angus, 124
McKinney, Louise, 26
McLaggan, J. W., 38
McLeod River 89
Me and Charlie Creek, 202
Métis, 49, 68, 157, 178
Miette Hot Springs, 55, **57,** 274
Mile 58 Cabin (Summit Cabin), 33, 147, 194, 196, 199, 205, 210, **211**
Milendorf, Charles E., 18, 26, 64, 292
Millar, H.S., 291
Millar, Willis N., **106**
Mills, J.H. (Jim), 82, 83, 84, 86, 96, 146, 292
Minnesota, USA, 126, 203
Moberly Creek, 18, 47, 71, 82, 124, 153, 169
Moberly Creek Lookout, 99
Moberly Ranger Station, **18, 210,** 230
Moberly, Alan, **276, 277**
Moberly, Ewan, 56, **68**
Moberly, Frank, **164, 165**
Moberly, Joe, **68**
Moberly, Lactappe, **68**

Moberly, Mabel, **276, 277**
Moberly, Madeleine, **68**
Moke the pack mule, **115**
Molstrom, P., 292
Monaghan, T. (Tom), **80, 81,** 134, 138, 152, **204,** 291
Monaghan's new store, 293
Montana Pete (Leatham, Jedidiah Jefferson), **15, 16**
Montana, USA, 30
Montezambert, Inspector, 10
Montreal Canadians, 270
Montreal Maroons, 270
Moose Creek *(see Moosehorn Creek)*, 48
Moosehorn Creek, 48
Morse code, 169
Morse, Charles H., 44, **107,** 192
Mount Bess, 34, **94,** 173
Mount Robson, 34, 119, 173, 181
Mount Sir Alexander, 173
Mountain District, 204
Mountain Trail, **13, 17, 18, 19, 22, 27, 33, 39,** 71, 96, 104, 147, 156, 158, 244, 257, 259, 273
Mounted Police *(see also RNWMP, RCMP)*, 64, 65, 185
Mouse Tower (Germany), 148
Muddywater River, 105, 113, 172, 202, 241, 250
Muddywater Valley, 109
Munro, George and Mrs., 290
Munson, Samuel (Sam), 27, 29, 30, 47, 59, 196
Murphy, Emily, 26
Muskeg Ranger Station, 155
Muskeg River Cabin, 65, **132,** 156, 217, 234
Muskeg River District, 238, 246, 274
Muskeg River, 36, 65, 132, 156, 158, **245**
Muskeg Trail, 155, 158, 160

N
Nail Keg Alley, 15
National Forestry Program, **266**
New Entrance, Alberta, 166, 204
New York City, 134
Neighbor, Rufe, 213, 290
Nickerson Creek, 87
Nickerson, Emma (Thorson), 86, **87**
Nickerson, Herman "Nick", 86
Nilson, Sam, **38**
Normandy, France, 49
North Battleford, Saskatchewan, 16, 65
North Saskatchewan River, 185
North Western Pulp & Power Ltd., 87, 107, 274
Nova Scotia, 175

O
O'Hagan, Dr. Thomas, 273, 274
O'Hagan, Howard, 15
Obed, Alberta, 253
Okanogan, 252
Old Entrance, Alberta, 16, 17, 135, 139, 204, 290
Old Trail (Lower Trail), 65
Oldman Creek Trail, 127
Oldman Creek, 127
Olson, Benjamin, 291
Onoway, Alberta, 98
Ontario, 119, 266
Orchard Creek, 124
Oregon, USA, 86
Ottawa Normal School, 14
Ottawa, 49, 106

P
Painter, Kenyon V., 119
Park Gate Station, **135**
Parlby, Irene, 26
Peace River, 84, 196, 233
Pembina River, 185
Persimmon Creek, 219
Pete, Montana, 64, 114, 117, 118
Pete's Flats, 16
Peterson, Mrs. Hans - (formerly Eileen Woodley), 290
Phillips Mountain, 119
Phillips, Donald (Curly), 118, **119, 190**, 221
Phillips, Harry, 73, 118, 155, 169, 219, **220,** 221
Pierre Grey's Lakes, **156, 157**
Pierre Gris, 157
Pincher Creek, 48, 49
Pinto Creek, 284
Plante, Caroline, 253
Plante, Felix, 43, 253, 292
Plushbottom, Lord, 124
Pocahontas, Alberta, 56, 139, 264
Powell, Russ, 107
Prairie Creek, 87, 114, 204
Prairie Creek ferry, **186**
Prince of Wales, 98
Prince of Wales Hotel, **46**
Prince Rupert, B.C., 86
Princess Creek, 55
Prohibition, 17
Prowse, Albert Henry (Bert), 290, 293
Public Works, Department of, 134
Pyramid Mountain, 120, 221

R
Rainbow Bear, 100, 101
Rance, Ranger, 235
Ranger School, 270, 271
Rawdon Township, Ontario, 14
Raymond, F., 292
Reader, L.H., 293
Red Pass Hotel, 139
Red Pass, B.C., 14, 79, 139
Red River, Manitoba, 178
Reimer, Abe, 182, 183, **293**
Reimer, Mary, 183
Regiment, 218th, 178
Rhein, Mr. R., 290
Rhine River, Germany, 148
Richardson, Walter L. (Walt), 293
Riel Rebellion, 16, 84
Rip Van Winkle, 36
Riviere, Henri "Frenchy" Arnous de la, **46,** 47, 48, **49, 50, 51,** 53, 56, 58, 63, 65, 71
Riviere, Henri, names of family: Alice (Primeau), Emily, Frances (Cox, later McWhirter), George, Henry, Inez (Rea), James, John "Charlie", Mary (Shriver), Nellie (Murphy), Robert, 49
Robson River, 119
Rock Creek Summit, 147
Rock Creek Valley, 147, 201, 256
Rock Creek, 33, 36, 211, 256, 258
Rock Lake Cabin, 22, 23, **130, 131, 205**, 259, **268, 269**,
Rock Lake, 16, 22, 23, 27, 29, 31, 62, 63, 71, 72, 113, 126, 147, 196, 203, 204, **244,** 250, **254, 255,** 257, 259, 260, 267, 270, 271, 272, 273, 278, 282
Rockslide Creek, 208
Rocky Mountains Park (Banff), 106
Rocky Mountains Forest Reserve, **52, 72,** 79, **99, 102, 103,** 115, **232**
Rocky Pass, **82**, 208
Rocky Pass Trail, 158, 160, 202
Rolla district, 89
Rome, Italy, 75
Royal Canadian Air Force, 266
Royal Canadian Mounted Police (RCMP), 48, 138, 204, 221
Royal North-West Mounted Police (RNWMP), 10, 15
Ruzicko, A.R.G., 292

S
Salter, W., 292
Savage .22 rifle, 105
Saw, Whip, **201**
Sawridge, Alberta, 14
Schmidt, Paul, **273**
Scidmore, E.C., 291
Second World War, 107
Senate of Canada, 26
Service, Robert, 90, 91
Seton Hospital, 274
Severson, O., 292
Shand Creek, 53
She Stoops to Conquer, 224

Sheep Creek, 72, 104, 105, **109, 174, 175, 181, 182, 183**, 202, 213, 264, 279, 282
Sheep Creek Pass, 169, 174
Sheep Creek Trail, 123, 169, 172, 184, 231, 282
Sheep Mountain, **82**
Shere, B.C., 91
Sherman, A., 293
Shuswap, Louie, 100, 101
Sibleyk, Jim, **276-277**
Silk Train, 16
Simonette River, 27, 30
Skookum House, 185
Skyline Trail, 114
Slave Lake Forest Reserve, pumps, **215**
Slave Lake, Alberta, 14
Smith, A.G., 293
Smith, William (Bill), 178, 196, **197,** 200, 201, 204, 205, 208, 210, 211, 217, 230, 293
Smith, J.M., 291
Smoky Mountain, 173
Smoky Range, **241**
Smoky River, 34, 106, 113, 119, 126, **150, 169, 170, 172, 175, 176, 177, 178, 184, 187, 189, 190, 191, 192,** 202, 213, **234, 239, 241**, 243, 247, 250, 282
Smoky River Cabin, 227, **228, 229**, 231, 232
Smoky River Fire, 238, 246
Smoky River Meadows, **71**
Smoky River Trail, 123, 184
Smoky River Valley, 173, 174
Snake Indian River, **55**, 232
Snake Indian River, Big Bend, **24, 25**
Snowfall, camping in, **110-111**
Soldier Settlement Board, Farm, 10, 78,
Solomon Creek, 16, 114, 244, 279
Solomon Creek Cabin, 114
Somme, Battle of the, 107
Speeder, railroad, Forestry, 136
Spriggs, Sergeant, 64, 65
St. Albert, Alberta, 157
St. Louis Flyers, 270
Stanton-Veale, Ralph, 139
Step and a Half, 138
Stiles, Sarah, 14
Stony River *(now Snake Indian River)*, 55
Styal, Alberta, 86
Sulphur Range, **82**
Sulphur River, 58, **60, 61, 148, 184,** 196, 199, 205, 211, 227, 239, 240, 243, **248, 249,** 250, 252, 279,
Sulphur River Valley, 33, 34, 39, 246
Summit Cabin (Mile 58), 196, 225, 233, 238, 250, 253, 256, 257, 267, 273, 278
Swedes (workers), 155
Switzer Provincial Park, 138
Switzerland, 49

T

Taylor, Dr. William C., 119
Teitge, F.W. (Fred), 293
Tête Jaune Cache, 16, 181
Thappe, Louise (Findlay), 84, 178
The Cattleman' 49
The Colt (horse), 27
The Compleat Angler, 126
The Deserted Village, 224
The Emperor Waltz, 165
The Flying Irishman, 213
The History of Little Goody Two Shoes, 224
The Vicar of Wakefield, 224
Thompson, David, 68, 84
Thompson, W.E., 291
Thoreau Creek (steam engine at), **114**
Tietge, Fred, **134, 245**
Timber Inspector, 226
Tonquin Valley, 114
Toronto, Ontario, 267
Totem Trophy, 165
Trail Riders of the Canadian Rockies, 114
Trojans, 143, 256
Truxler, Jackie, **276-277**
Truxler, Mark, 119, **254, 255, 273**, 274
Truxler, Mark and Mrs. (Agnes), 290
Truxler, Vernon (Vern), 274, **276-277**

U

U.S. army .30 calibre (rifle), 105
U.S. Forest Service, 106
Udell, Bob, 107
Unemployment Relief Camps (URC), 266
University of Toronto, forestry, 106, 107, 291
Upper Sulphur River, 207
Utah, USA, 15, 16

V

Vancouver Island, B.C., 224
Vancouver, B.C., 139
Vernon, B.C., 224, 238, 252
Victoria, B.C., 14

W

Waikle, L.L., 262, 284, 293
Wallingford, Get-Rich-Quick, 74
Walton Creek Trail, 279
Walton, Isaac, 126, 127
Wanyandie Flats, 138
Warner, J.W., 291
Washington, USA, 252
Waterton Lakes National Park, **46**
Watt, Andrew Gordon, (Gordon), 106, **150,** 278, 293
Webb, Dora, 139
Welch, T.H., 291
Wells, W.C., 292
Weorzba, William, 292
Wharton, Francis (Fred), 259, 260, 261
Wheeler, J.E., 292
Whimster, H.M., 292
Whispering Cliff (Faulk), 202
Wilderness Men, 15
Wildhay River, 16, 22, 23, 29, 33, 34, 36, 62, 82, **100,** 101, 114, 154, 155, 169, 258, 270, 272, 282
Wildhay River Cabin, 96, 99, **265**
Wildhay-Moberly Trail, 155
Wilmore Wilderness Park, **58,** 114
Wilson, Geoffrey, 221
Winnifred Camp, 58
Winnifred Lakes, **58**
Winter Creek, 18, 22, 99, 118, 122, 123, 153, **154,** 166, **167,** 169, 265, 282
Women are Persons statue, 26
Woodley Bros. General Merchants, **73,** 139
Woodley Brothers, Earl and Roy, 139
Woodley family, 139
Woodley Schoolhouse, 136, **137**
Woodley, Earl and Edna, **139**
Woodley, Earl, 14, 79, 139
Woodley, Miss Mildred, 294
Woodley, Mr., buggy, 153
Woodley, R.T. (Roy), General Merchant, 14, 139, 204
Woodley, Roy and Mrs. (Edna), 294
Wright, Jack, 107
Wright, R.G., 293
Wroe Creek, 284
Wroe, Percy, 293

Y

Yale University, 106
Yellowhead Pass, 16
Yim and Yeff, packhorses, 27
Yorkton, Saskatchewan, 14
Youth Forestry Training Program (YFTP), 266
Yukon Gold Rush, 16